Cambridge Elements

Elements in the Philosophy of Physics
edited by
James Owen Weatherall
University of California, Irvine

PHILOSOPHY OF PHYSICAL MAGNITUDES

Niels C. M. Martens
Utrecht University

Shaftesbury Road, Cambridge CB2 8EA, United Kingdom

One Liberty Plaza, 20th Floor, New York, NY 10006, USA

477 Williamstown Road, Port Melbourne, VIC 3207, Australia

314–321, 3rd Floor, Plot 3, Splendor Forum, Jasola District Centre, New Delhi – 110025, India

103 Penang Road, #05–06/07, Visioncrest Commercial, Singapore 238467

Cambridge University Press is part of Cambridge University Press & Assessment, a department of the University of Cambridge.

We share the University's mission to contribute to society through the pursuit of education, learning and research at the highest international levels of excellence.

www.cambridge.org
Information on this title: www.cambridge.org/9781009467827

DOI: 10.1017/9781009233705

First published 2024

A catalogue record for this publication is available from the British Library.

ISBN 978-1-009-46782-7 Hardback
ISBN 978-1-009-23368-2 Paperback
ISSN 2632-413X (online)
ISSN 2632-4121 (print)

Philosophy of Physical Magnitudes

Elements in the Philosophy of Physics

DOI: 10.1017/9781009233705
First published online: February 2024

Niels C. M. Martens
Utrecht University

Author for correspondence: Niels C. M. Martens, martensniels@gmail.com

Abstract: Dimensional quantities such as length, mass and charge – that is, numbers combined with a conventional unit – are essential components of theories in the sciences, especially physics, chemistry and biology. Do they represent a world with absolute physical magnitudes, or are they merely magnitude ratios in disguise? Would we notice a difference if all the distances or charges in the world suddenly doubled? These central questions of this Element are illustrated by imagining how one would convey the meaning of a kilogram to aliens if one were allowed to communicate only via Morse code.

This Element also has a video abstract: www.Cambridge.org/EPPH_Martens

Keywords: physical magnitudes, physical quantities, scaling symmetries, metaphysics of quantities, mass

ISBNs: 9781009467827 (HB), 9781009233682 (PB), 9781009233705 (OC)
ISSNs: 2632-413X (online), 2632-4121 (print)

Contents

1 Introduction

Philosophers often operate with a misleading caricature of scientific statements in terms of propositions from first-order logic with *primitive*, *qualitative* predicates, such as 'For all x, if Rx then Bx', where R and B are to be understood as, for example, 'being a raven' and 'being black', respectively. The much richer properties that are in fact central to sciences such as physics, chemistry and biology are physical magnitudes represented by numerical quantities – for example spatial length, temporal duration, electric charge and mass – and/or relations between these magnitudes. When planets and molecules 'have mass' or 'are spatially separated', they instantiate these determinables in virtue of instantiating determinate physical magnitudes or determinate magnitude ratios, each with a rich quantitative structure.[1] For example, that the Earth and Mars are massive is true in virtue of them standing in a determinate mass relationship, namely a ratio of 1:0.107. A common claim is that an electron is charged because it has a determinate electric charge that we refer to as $1.6 \cdot 10^{-19}$ Coulomb. Moreover, most of these physical magnitudes are dimensional: that is, their representation in terms of numerical quantities depends on a conventional unit. For instance, until recently the masses of objects were expressed in terms of their relationship to an arbitrarily chosen object, stored in Paris, that served as the standard unit of mass.

This Element focuses on the question of how we should understand the dimensional physical determinables – and their associated determinates, whether these are absolute (i.e. monadic) magnitudes or magnitude relations – that are central to many of the natural sciences, in a way that does justice to these sciences. What is the world like – that is, what metaphysics corresponds to a world such as ours that is described by dimensional physical determinables?

So-called Ozma problems or Ozma games provide a particularly vivid and useful illustration of the various sub-questions and nuances at play. Ozma games, named after Project Ozma, which searched for radio signals from extraterrestrials, challenge us to convey the meaning of various physical properties to aliens, solely by sending them a signal in Morse code, without being allowed to ostensively refer to (e.g. point at) objects in the night sky that instantiate such properties. For many properties faithfully represented by a dimensionless integer *n*, such as the number of corners of a square or the number of fingers on a hand, this is easy: simply send them *n* beeps, which they can count. More challenging is Gardner's (1964/1990) original Ozma problem, which concerns

[1] See Wilson (2012) for a criticism of the determinable/determinate model, and Wolff (2020, chapter 2) for a criticism of its application to quantities/magnitudes. Note, however, that nothing in this Element crucially depends on this model.

the properties of 'left-handedness' and 'right-handedness'. Trying to communicate 'left-handedness' by operationally defining it in terms of the majority of the mass of a human heart being on the left side of our body will not succeed, as this is a contingent feature; the laws of nature allow that the aliens, if they have a heart at all, have theirs skewed towards the right side of their body. The same applies to the right-handedness of our DNA's double helix. To solve the problem, we have to instruct the aliens to perform an experiment on a system that is not governed by mirror-symmetric laws. Fortunately, (only) one of the four fundamental interactions of nature, the weak interaction or weak nuclear force, violates mirror-symmetry. This is the force that governs the radioactive decay of atoms. So we could instruct the aliens to observe the handedness of the decay process of, say, Cobalt-60, telling them that that asymmetry defines (what we conventionally call) left-handedness.[2]

Is it possible to convey the full meaning of dimensional determinables such as length, charge and mass to the aliens – that is, including any and all determinates in virtue of which such a determinable obtains? If so, how? In this Element our main example will be (Newtonian) mass,[3] although most considerations carry over to other determinables (see for instance Sebens (2021, section 4.2) for a charge analog of the argument in Section 2.2 of this Element). One may have thought that the Ozma mass challenge translates into – fully reduces to – the task of having the aliens pick out an object in their neighbourhood of which we would say that it has a mass of, say, 1 kg, or, equivalently, sending them a 1 kg object and enquiring after its properties. The answer is easy: describe an electron, specify the mass of an electron in kilograms, and determine the mass ratio between an electron and the object in question. However, as we will see in the remainder of this Element, the fascinating metaphysical and conceptual subtleties of dimensional determinables become much clearer if we only allow ourselves tools from the regime in which Newtonian gravity is approximately empirically adequate. With this restriction, if the aliens would be able to express the object's determinate mass at all, it would presumably be expressed as something like 70,863 quohrts. Upon further prodding, they might explain that the object we sent them just happens to stand in a mass ratio of 70,863:1 with respect to the standard object of 1 quohrt that is safely stored in their capital city. This reveals the defining aspect of dimensional determinables, which one

[2] More precisely, it is your left and not your right hand for which it holds that, if you curl the fingers of your hand to follow the spin, the decay product will predominantly follow the direction of your thumb.

[3] Since the weak equivalence principle holds, we do not have to distinguish here between inertial and gravitational mass.

may refer to interchangeably as the kinematic redundancy or absolute indistinguishability of absolute magnitudes, the representational indeterminacy or arbitrariness of absolute magnitudes, or the conventionality of units.[4]

> **Representational indeterminacy of absolute magnitudes:** For any dimensional determinable, such as mass, the monadic (i.e. 'one-place') magnitude predicated of any particle can be reported or expressed, non-dynamically, only in terms of how this magnitude *relates* to the magnitude of another particle having the same determinable property. Equivalently, representing these magnitudes requires both a numerical quantity and a unit; since the choice of the latter is conventional, there is no unique numerical quantity that represents a specific magnitude.[5]

Some would argue that this is the case because there are in fact no absolute magnitudes; the concept of mass – the dimensional determinable mass – is exhaustively described in terms of a(n infinite) set of determinate mass relations. (Here we will follow Baker (manuscript, May 2013) in focusing only on scale-invariant mass relations: that is, those that are to be numerically represented by ratios. I will often refer to those simply as 'mass ratios'.) Those who argue for this are realists about mass ratios, but anti-realists about absolute masses; that is, they do not consider the latter to be merely kinematically redundant, but fully, metaphysically redundant. This full-blown anti-realism about absolute magnitudes will be the main view of contention in this Element.

> **(Anti-)realism about absolute magnitudes:** There (do not) exist absolute (i.e. monadic)[6] physical magnitudes, such as absolute masses, that are predicated of – had by – some of the objects in the theory.

On this view, phrases such as '1 kg' are mistakenly interpreted as being a one-place property ascribed to a single object; they are actually two-place properties, namely ratios, in disguise, as 'kg' is best understood as referring to another object – in the recent past this was the standard kilogram in Paris. One of the main questions, when trying to understand what the world is like given that it is described by theories with dimensional determinables, is thus whether this requires absolute magnitudes in virtue of which the objects instantiate that

---

[4] See Jacobs (2021), section 6.6, for a critical discussion of the closely related notion of 'kinematic comparativism'.

[5] Compare Principle III of Peacocke (2019, chapter 2).

[6] In the case of, for example, distances, the determinate magnitudes (that one is to be realist or anti-realist about) would be dyadic (i.e. two-place) relations. Absolutism and comparativism (see main text) regarding distances would then concern four-place comparisons/relations, for example the distance between particles 1 and 2 being twice the distance between particles 3 and 4.

determinable, or not. Note that the answer may differ for each determinable (in each theory).

If the concept of mass (in a given theory) is exhausted by dimensionless mass ratios, not requiring any absolute, dimensional masses, then the Ozma mass problem is almost as simple as communicating the number of fingers on a hand. With equipment such as balances, the aliens could determine all the mass ratios of objects in their neigborhood. If we sent them a 1 kg object they could also determine all mass relations with respect to that object. If we did not, then it seems that they would not be able to pick out an object that we would refer to as having a mass of 1 kg, but that would only mean that they are unaware of our conventions and of relations between a pair of objects, one of which is in their neighbourhood and the other in ours. They would lack no local understanding of the concept of mass whatsoever. It seems that we would have been able to communicate everything there is to communicate about (the objective and local aspects of) the concept of mass.

However, anti-realism about absolute masses does not follow simply from the conventionality of units. Absolute masses would need to be redundant in all possible ways. Although there is no intrinsic difference between left-handed and right-handed objects – they agree on all the relations between their own parts – one of the laws of nature has nevertheless ensured an empirical difference associated with handedness, such that we could explain the concept of left-handedness to the aliens in terms of an experiment. Similarly, although absolute masses are not absolutely discernable (see also Section 3) – which is the precondition for the conventionality of units – a question remains as to whether absolute masses are empirically relevant. Three different aspects of the Ozma mass problem are to be distinguished (Martens, 2021):

Ozma mass subquestion 1 [kinematic relevance] If there are absolute masses, could we express a specific absolute mass (say the one that we have labelled '1 kg') to the aliens in the same direct (i.e. kinematic) sense as we can convey dimensionless determinates?

Ozma mass subquestion 2 [empirical relevance] Could the aliens determine (e.g., detect/ discover) that absolute masses are empirically relevant?

2a [dynamic relevance] Could the aliens do an experiment to determine that varying absolute masses while leaving the mass relations unchanged leads to an empirical difference?

2b [explanatory relevance] Are there other empirical, observable facts that require postulating absolute masses to be fully explained?

Ozma mass subquestion 3 [indicative determinacy] Can we ensure that the aliens pick out a specific object with a mass that is the same as the mass that we have labelled, say, '1 kg'?

We have seen that the answer to subquestion 1 is 'no': absolute masses are kinematically redundant. I will only briefly return to subquestion 3 in Section 8. We will see in Section 2 of this Element that subquestion 2 is crucial for settling the debate between realists and anti-realists about absolute masses.

Much of the literature focuses instead on a related but distinct dichotomy, concerning the relative fundamentality of mass ratios and absolute masses:[7]

> **Absolutism:** There exist determinate absolute masses in virtue of which (facts about) determinate mass relations obtain, or, equivalently, absolute masses are more fundamental than mass relations.
>
> **Comparativism:** The denial of absolutism: determinate mass relations do not obtain in virtue of (more fundamental) absolute masses – if the latter exist at all.

Discussions or adamant defenses of comparativism can be found in Russell (1903), Ellis (1966), Field (1980), Bigelow et al. (1988), Bigelow and Pargetter (1990), Arntzenius (2012), Dasgupta (2013) – it was Dasgupta who first coined the terms absolutism and comparativism in this context – Eddon (2013), Perry (2016), Roberts (2016) and Jalloh (2022). Absolutism has been defended in one form or another by Armstrong (1978, 1988), Lewis (1986) – it is embedded in the standard Lewisian version of Humean supervenience (see Section 6.2), which postulates a spatiotemporal mosaic with monadic masses and charges sprinkled all over – Mundy (1987) and Jacobs (2023b). The views of Wolff (2020) and Dewar (2020) are sometimes claimed to be a third position, 'sophisticated substantivalism', but in this Element they are understood as being a species of absolutism, so-called sophisticated absolutism (see Section 3 and Dewar (2020, fn.10)).

It may seem that the realism/anti-realism and absolutism/comparativism dichotomies coincide. This would indeed be the case if the only way for absolute masses and mass ratios to coexist were such that the mass ratios obtain in virtue of the absolute masses – after all, (the numerical representations of) mass ratios are standardly conceived of as the ratio of the numerical quantities representing the masses of the two relata. If so, comparativism would always imply anti-realism about absolute masses. While this is indeed very much in the spirit of comparativism, we will see in Section 6.2 that there is a (Humean) view that is strictly speaking a form of comparativism, even though it does acknowledge the reality of absolute masses – see Figure 1. In light of this, and to avoid to some extent the controversial notions of 'relative fundamentality' and especially the related notion of 'grounding', the realism issue will take

[7] In Martens (2021) these positions are referred to as weak absolutism and comparativism, as they concern relative rather than absolute fundamentality.

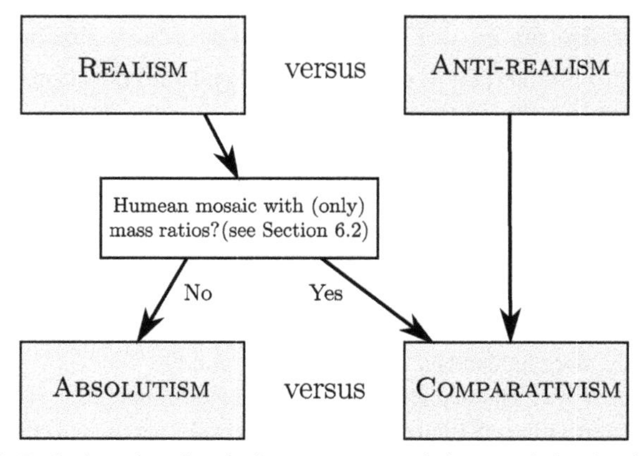

Figure 1 Relating the absolutism–comparativism and the (anti-)realism debates

centre stage in this Element. Finally, it is important to note that it is consistent to be a realist about absolute magnitudes of some determinables but not others, just as it is consistent to be an absolutist about some but a comparativist about others.

What follows is perhaps best understood as a recipe book for constructing one's preferred ontology for physical regimes in our universe which are accurately described by theories that feature dimensional determinables: that is, all current theories in physics (and many aspects of chemistry and biology). The considerations revealed by the Ozma challenge will be formalised in Section 2. Each subsequent section concerns a decision point: whether to add or purposely leave out a specific ontological ingredient, with an explanation of the benefits and drawbacks to each choice – this decision tree is represented in Figure 2.

2 Arguments

2.1 Realism, Anti-realism and Agnosticism

We have seen that the conventionality of units does not suffice to vindicate anti-realism about absolute magnitudes (even though this reasoning seems to be implicit in, for instance, Mach (Martens, 2018) and Roberts (2016)). The crux is whether absolute masses are or aren't empirically relevant. The typical main argument for anti-realism about absolute masses is therefore similar to (modern versions of) arguments for anti-realism about absolute positions, absolute velocities and absolute space that originate with Newton and Leibniz (Alexander, 1956/1717):[8]

[8] See Dasgupta (2013) for a discussion of various other arguments.

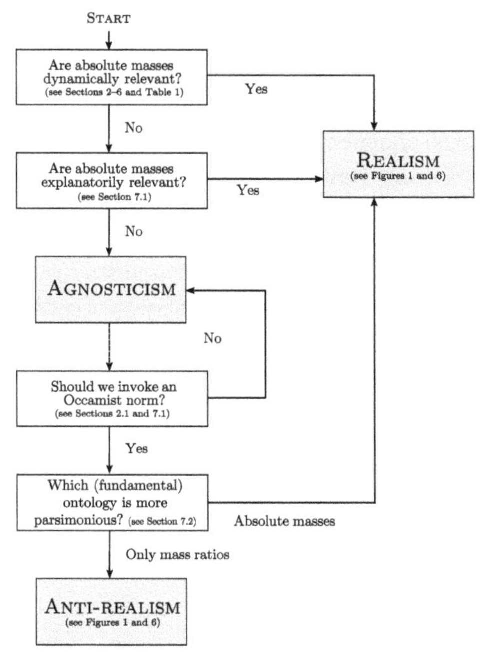

Figure 2 Decision tree regarding realism, agnosticism and anti-realism about absolute masses

---

P_1 **Dynamical irrelevance:** Absolute masses are dynamically irrelevant.

P_2 **Explanatory irrelevance:** Absolute masses are explanatorily irrelevant.

P_3 **Occam's razor:** All other things being equal (i.e. absolute masses being empirically – both dynamically and explanatorily – irrelevant), the more parsimonious ontology is the correct one.

P_4 **Ontological expensiveness:** An ontology of only mass ratios is more parsimonious than an ontology of absolute masses and mass ratios.

C Anti-realism about absolute masses.

Occamist norms are thrown around often in the literature, without it always being made clear what the content and the status are of such principles. The content will be discussed in Section 7.2, which will require a revision of the anti-realism argument. Regarding the status: do Occamist norms convey an absolute truth? Is it impossible or even incoherent for the world to contain entities that are unobservable, irrelevant, inert? Although the logical empiricists thought so, with their verifiability criterion of meaning, the modern answer tends to be 'no'. Moreover, absence of evidence (for absolute masses) is not evidence of absence. Perhaps Occam's razor is then best understood in a weaker sense, as a methodological principle of epistemic humility, changing the focus of this argument to an argument against realism about absolute masses without taking a stance on anti-realism about masses (see Figure 2):

P_1	**Dynamical irrelevance:** Absolute masses are dynamically irrelevant.
P_2	**Explanatory irrelevance:** Absolute masses are explanatorily irrelevant.
P_3	**Occamist humility:** Do not commit to the existence of some entity without proper justification, even if one cannot justify the denial of its existence.
C	Agnosticism about absolute masses.

Since much of the literature includes parsimony considerations, we will, for the sake of completeness, discuss these in Section 7.2. Explanatory considerations will be covered in Section 7.1. However, dynamical (ir)relevance, which features in both versions of the argument, will form the primary focus of this discussion, starting in this section and also underlying much of Sections 3–6.

2.2 Dynamical (Ir)relevance

Are absolute masses dynamically relevant? Can Newtonian gravity with an ontology that excludes absolute masses be empirically adequate (in the Newtonian regime)? Is Newtonian gravity with absolute masses empirically equivalent to Newtonian gravity with only mass ratios? Could the aliens do an experiment to determine that varying absolute masses leads to an empirical – that is, observable – difference?

According to Dasgupta (2013), who is a comparativist and anti-realist about absolute masses, it is physically impossible for a device to exist that would detect absolute masses, in the sense of (a) providing one specific output, for instance displaying '1 kg' on a computer screen, if and only if it was presented with a 1 kg object; and (b) providing another specific output, for instance displaying '2 kg', if and only if it was presented with a 2 kg object. The realist and anti-realist about absolute masses disagree about whether an observable difference would result from the following operation (which would leave all the mass ratios unchanged):

> **(Exclusive) active Leibniz mass scaling:** A uniform scalar multiplication of the absolute mass magnitudes of all of the massive objects in the universe, *ceteris paribus* (i.e. while keeping everything else that is independent from the masses unchanged).

'Exclusive' refers to the strict *ceteris paribus* clause, which insists that when trying to understand the concept of mass you need to vary only masses and nothing else (unless it depends on mass, see Section 5), not even dimensional constants such as Newton's gravitational constant G.[9] We will omit mentioning the 'exclusive' qualifier until we explicitly consider inclusive scalings that do scale G along with the masses, in Section 6.3. 'Active' refers to the magnitudes 'out there in the world' being scaled and not merely the unit or the numerical quantities representing the magnitudes being scaled. 'Leibniz' is used by analogy with Leibniz shifts – for example, moving all the matter in the universe three meters to the east – in the debate between Newton, Clarke and Leibniz about absolute and relative motion and the fundamentality of space; it concerns a uniform scaling of a whole (possible) world, rather than of only a subsystem of a world. According to a realist about absolute masses (who is motivated by the dynamical relevance of absolute masses), such a scaling would lead to an observable difference. For the anti-realist or agnostic – although this scaling is strictly speaking ill-defined for the anti-realist – this transformation must effectively be the identity operation. Nothing changes. Thus, if the hypothetical detection device would display '1 kg' when presented with a 1 kg object, in a mass-doubled world it must still display '1 kg' on the observable computer screen even though it is in this case presented with a 2 kg object (thereby violating Dasgupta's condition b – and a). Hence, the device would not actually be an absolute-mass-detecting device; it just got lucky in the first of the two possible worlds.

One way for the realist to argue for the dynamical relevance of absolute masses is via the following thought experiment, which captures the essence of Baker's Earth-Pandora argument (Baker, manuscript, May 2013). Consider a two-particle world governed by the laws of Newtonian gravity. (How we are to understand these laws is a tricky question to which we will return in Section 6.) Suppose, without loss of generality, that the particles are equally massive: that is, they stand in a 1:1 mass ratio. Suppose moreover that initially they are a distance r_0 apart and are moving rectilinearly away from each other with relative velocity v_0 (such that there is zero angular momentum, to keep things simple). This scenario is depicted in Figure 3. What will happen?

[9] The 'exclusive' terminology in this context was coined by Jacobs (2023b).

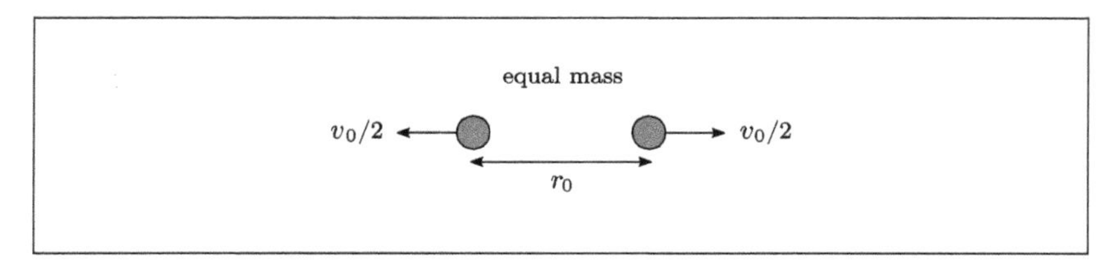

Figure 3 Two-particle scenario

Using the standard formulation of Newton's law of universal gravitation, $F_g = \frac{GM_1M_2}{r^2}$, and Newton's second law, $F = Ma$, one can derive the following escape velocity inequality:

$$v_e = \sqrt{\frac{2GM}{r_0}}. \tag{1}$$

If the initial relative velocity v_0 is larger than v_e it will escape in finite time; if it is equal it will escape in infinite time; and if it is smaller than v_e the gravitational pull will be too large and provide enough of a brake for the particles to 'collide': that is, to eventually coincide. These two types of evolution – escape and coincidence – are clearly observationally distinct; more on this in the paragraphs that follow. Crucially, v_e is a function of the absolute mass, rather than (just) mass ratios. The realist claims that only specifying the mass ratios, in this case 1:1, is insufficient to determine the evolution of the system. If in one instance of this thought experiment the absolute masses M are such that the velocity v_0 is just smaller than the escape velocity, this will produce a coincidence evolution like the one in Figure 4(a). If one now considers another possible world, differing from the previous one only by the masses being halved, $M \to M/2$ (i.e. r_0 and v_0 are kept constant; we'll return to this in Section 5), the evolution will be observationally distinct: an escape evolution like the one in Figure 4(b). As far as the anti-realist about masses is concerned, the initial conditions of both of these possible worlds are identical, so their subsequent evolution should be indistinguishable.

Note that the realist's case does not depend on the exotic, idealised, artificial two-particle scenario. As this is the only scenario that we can deal with analytically, it is pedagogically favoured. The argument does generalise to multi-particle scenarios, although they require numerical methods to be solved. This is illustrated with two three-particle examples in Figure 5.

Let us briefly return to the claim that escape and coincidence evolutions are observationally distinct. One way of seeing this – in the two-particle scenarios – is that in escape scenarios the distance between the particles keeps increasing over time, or in other words, the ratio of a later to an earlier distance

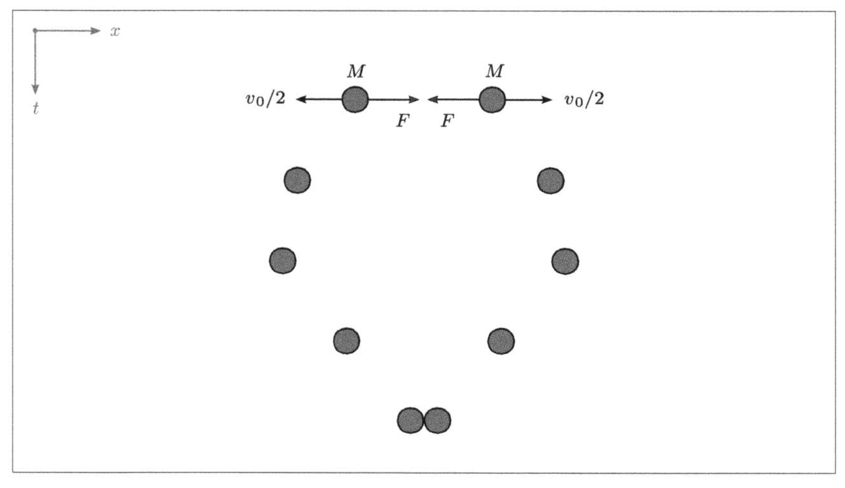

(a) Coincidence scenario

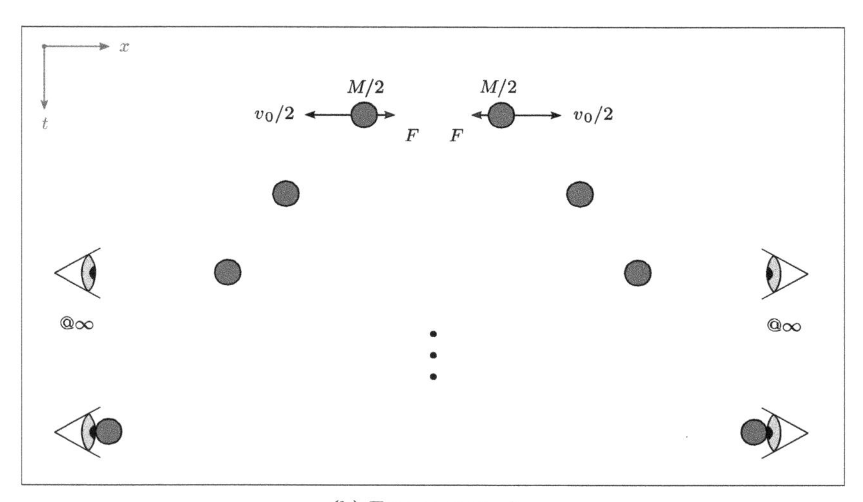

(b) Escape scenario

Figure 4 Two categories of possible evolutions of the (incomplete) initial state in Figure 3, according to the realist about absolute masses

is always larger than 1, with the opposite being the case for collision scenarios. This difference should be recognisable by both the realist and the anti-realist. However, since there are, strictly speaking, no rigid rods available in Newtonian gravity, one may also describe the difference purely topologically: in two-particle escape solutions the particles will never coincide – that is, their trajectories will never intersect, which, as the name suggests, is not the case for coincidence solutions (even though the theory will break down at that very point). (In the three-particle scenarios in Figure 5 we notice that in the solid-line

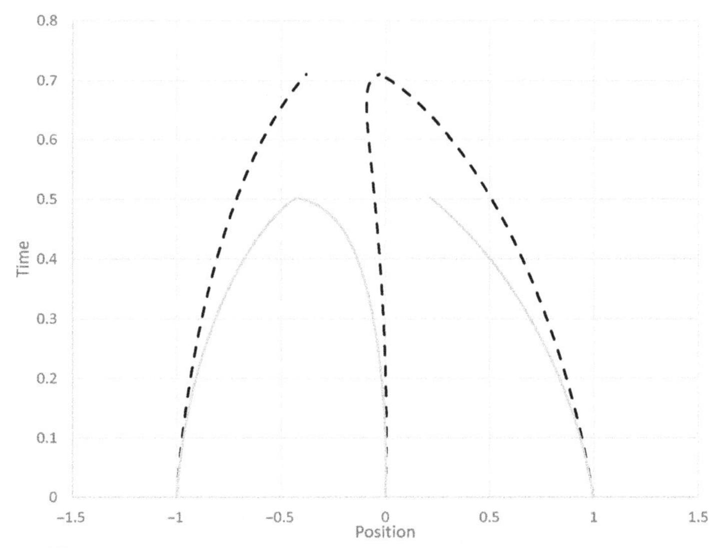

(a) Both three-body solutions: $G = 1$, $r_{12} = r_{23} = 1$, $v_1 = 0.2$, $v_2 = 0.1$, $v_3 = -0.5$. Dashed-line solution: $m_1^d = 3$, $m_2^d = 0.75$, $m_3^d = 2$. Solid-line solution: $m_i^s = 2m_i^d$.

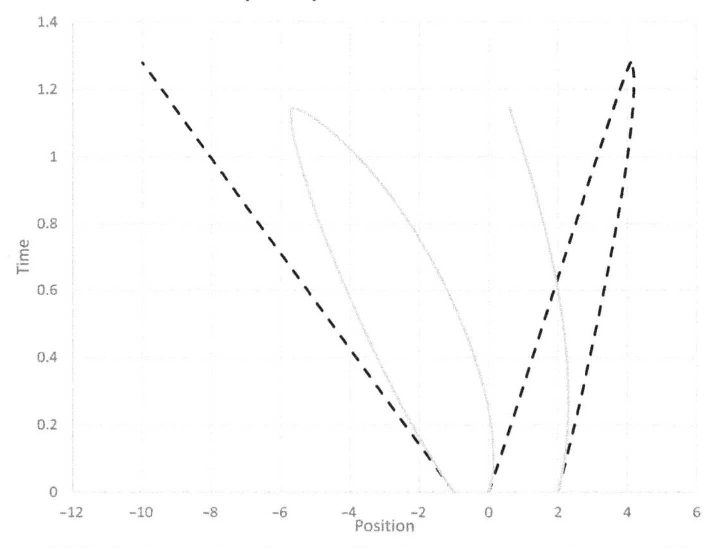

(b) Both three-body solutions: $G = 1$, $r_{12} = 1$, $r_{23} = 2$, $v_1 = -7.2$, $v_2 = 3.5$, $v_3 = 2.5$. Dashed-line solution: $m_1^d = 6$, $m_2^d = 2$, $m_3^d = 0.5$. Solid-line solution: $m_i^s = 11m_i^d$.

Figure 5 Three-particle scenarios. In each frame, two separate numerical solutions of the three-body problem have been superimposed. The initial states of both solutions agree on the distances, velocities and mass ratios, but the absolute masses differ by an exclusive active Leibniz mass scaling. Each three-body problem has only been solved until the first collision, as the theory breaks down at that point. Subscripts refer to particle number.

solutions the two particles on the left coincide (first) whereas in the dashed-line solutions it is the two particles on the right that coincide (first).) Moreover, if we allow ourselves an observer, or a detector (with sufficiently low mass as to not disturb the two-particle system), or even just an imaginary line, we can in fact create the absolute mass detection device that Dasgupta claimed was physically impossible. Place an observer or detector at infinity, which outputs '$M < M_{esc}$' if and only if the particle reaches the location of the observer/detector: see Figure 4(b). (Or place the detector at various smaller, finite distances, to determine whether a larger mass boundary is crossed or not.)

Are we to agree with this story as told by the realist? This depends on how one thinks about various implicit assumptions that were being made. How are we to understand laws of nature such as the laws of Newtonian gravity? Do they (indeed) require reference to absolute magnitudes? And how are we to understand the behaviour of dimensional constants appearing in the laws when scaling magnitudes with a dimension shared by that constant (Section 6)? Moreover, why is it r and v that are to be kept constant when uniformly scaling the masses (Section 5)? These decision points are summarised in Table 1. Before turning to such issues, the next section provides a range of possible realist and anti-realist ontologies to choose from. Even though this is the endpoint of the journey that we will embark upon in the next sections – see Figure 2 – some knowledge of the destination will help to guide us on each of the individual paths.

3 Decision Point: Physical Magnitudes

The ultimate decision to be made is the precise ontology of absolute masses and/or mass relations. We start here by providing a range of options in this regard – depicted in Figure 6 – although the final decision can only be made after all the other decision points in this Element have been considered (Figure 2).

Let us start with absolute magnitudes, as an absolutist might construe them.[10] They comprise a set of monadic (i.e. 'one-place') relata (we will discuss the choice of relata and the qualifier 'monadic' later in this section). We start by assuming that this set has cardinality $\aleph_1$ (but see below), to make sure that

[10] This definition of absolute magnitudes will suffice for many typical examples, such as mass, charge and length, but it may not work for all magnitudes. Caspar Jacobs suggested temperature on an interval scale (for instance in Celsius) and a U(1) gauge field as potential counterexamples (a personal communication) (My personal view is that the temperature counterexample forms no problem, as it boils down to a suboptimal choice of representation in terms of numerical quantities, which is avoided if one considers temperature in Kelvin.)

Table 1 Steps in determining the dynamical relevance of absolute magnitudes such as absolute masses (Step 5 is not covered in this Element but is detailed in the cited works)

Step 1	Determine the full set of **independent initial variables/determinables**, and focus on the absolute magnitudes of one (dimensional) determinable (e.g. mass) of which we are considering the appropriate ontology (Section 5).
Step 2	Determine the appropriate (form and interpretation of the) **laws/dynamics** (Section 6).
Step 3	If these laws include a **dimensional constant of nature** (for which the dimension of the determinable of interest is non-zero), should we scale that constant along when scaling the absolute magnitudes of interest, or should we keep it constant (Section 6.3)?
Step 4	All else being equal (i.e. steps 1–3), does an **active Leibniz scaling** of the absolute magnitudes of interest lead to an observationally distinct possible world / is such a scaling undetectable / is such a scaling a symmetry (Section 4)? If so, these absolute magnitudes are dynamically relevant (Section 2.2 and Ozma subquestion 2a in Section 1).
(Step 5)	(When considering all possible scalings/determinates of all initial variables, do we recover – up to observational distinctness – all the possible worlds that we are suppose to recover (**completeness**) and do we avoid possible worlds that are not supposed to be dynamically possible (**soundness**) (Section 4; Sections 6.1, 6.3, 6.4; Martens (2022, section 4); Martens (2017a, chapter 3))?)

we include – in terms of the numerical quantities that can be used to represent them – all the possible finite, positive mass values, or, in the case of electric charge, all the possible finite charge values. Two structures are imposed upon the elements of this set: 1) a total order, and 2) an associative concatenation structure. The operational interpretation of these structures can be illustrated with the case of mass. The first structure corresponds, for instance, to comparing the masses of two objects by placing these objects on the opposite scales of a balance. Adding the second structure corresponds to allowing multiple massive objects on each scale of the balance, and comparing the mass of the two collections/concatenations. This second structure fixes where the concatenation of several magnitudes 'fits into the total order'. (Once we represent these magnitudes by numbers we will be able to interpret the second structure in addition.) If the concatenation of each pair of elements is also included in our set of absolute magnitudes (which motivates our choice of cardinality $\aleph_1$), called 'closure', this set together with its two structures comprises either

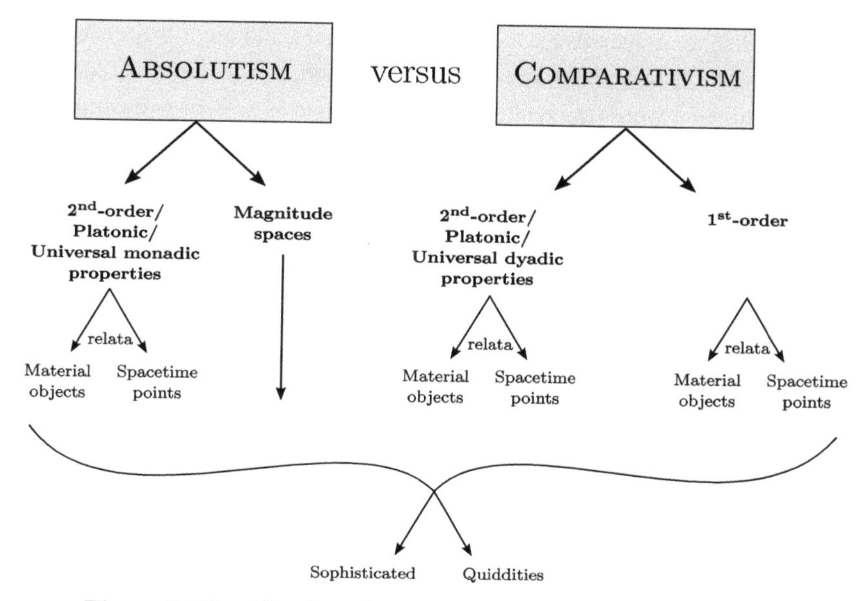

Figure 6 Classification of the main ontologies to be chosen from

a totally ordered semi-group (for the case of mass, where there is no inverse, and if there is no zero charge no identity element either) or a totally ordered group (in the case of electric charge, as negative charges provide inverses to the positive charges and the neutral charge constitutes the identity element). Representation theorems then show that structures like these can be faithfully represented by $\mathbb{R}$ (for electric charges) or $\mathbb{R}^+$ (for 'non-zero', 'positive' masses), and uniqueness theorems show that these representations are unique up to a constant: a choice of unit. One way of seeing how (facts about) mass ratios will then obtain for the absolutist – that is, in virtue of these more fundamental absolute masses – is at the level of representation: from any numerical representation of the absolute masses one can generate the same unique numerical representation of the scale-invariant mass relations simply by dividing the two numerical quantities representing the absolute masses of a pair of particles to obtain the corresponding mass ratio.

Now consider scale-invariant mass relations, that is, mass 'ratios', as a comparativist might construe them. Perhaps surprisingly, they have a structure that is rather similar to that of the absolute magnitudes. One crucial difference is that they comprise a set of *dyadic* (i.e. 'two-place') relata. Nevertheless, the standard assumption for this set of relata is once again that it has cardinality $\aleph_1$, this time to ensure that we include all the possible positive mass ratios. As before, two structures are imposed upon the elements of this set: 1) a total order, and 2) an associative concatenation structure, although this concatenation structure is to be interpreted as multiplication.[11] It includes the identity element, in other words the relation '... is as massive as ...' – or in quantitative terms '... stands in a 1:1 mass ratio to ...' – and also includes an inverse for every mass ratio. Hence, the mass ratios form a totally ordered group.

What exactly are the relata – the elements of these two sets of absolute masses and mass ratios? On one typical construal they are determinate *properties* that fall under a single determinable (e.g. mass). (Sometimes these are referred to as 'universals'.) These construals tend to be referred to as 'second-order', since one may describe them in terms of second-order logic, because the ordering and concatenation structure are relations not between zeroth-order objects but between first-order properties. Mundy (1987) famously defends absolutism in this second-order form. Now, the difference between (absolutist) absolute masses and (comparativist) mass relations is that the former are monadic properties and the latter are dyadic properties (i.e. relations). In either case, these properties could be instantiated by (one or a pair of) material objects or by (one or a pair of) spacetime points. The latter option sometimes goes under the names of supersubstantivalism or monistic substantivalism.

For the absolutist there is an alternative ontology of absolute masses, and for the comparativist there is yet another alternative ontology of mass ratios (Figure 6). Let us start with the latter. Those who dislike Platonic/second-order ontologies may wish for the two structures to be instantiated directly by objects instead of being mediated by properties – that is, the first-order option. Again, these objects could be either material objects or spacetime points. Field (1980) famously defended a verison of comparativism with 'quantitative' structures (albeit different ones from the two mentioned above) instantiated directly by spacetime points. Second-order advocates are sometimes motivated by their intuition that statements such as 'the mass of a is half the mass of b' have as their natural interpretation that one is comparing properties of the objects a and b rather than directly comparing the objects themselves (Mundy, 1987, p. 33–34). However, it is difficult to know what to make of this since others have the opposite intuition: that one is in fact ordering the objects themselves rather than any of their properties (Wolff, 2020, p.125). Perhaps more serious is the following objection, raised for instance by Mundy (1987) against first-order views such as those of Nagel (1932), Krantz et al. (1971), Roberts (1979) and Field (1980). The closure condition for the (semi-)group structures mentioned above is an (implicit) existence condition/postulate: for each pair of elements

[11] These structures correspond nicely to Weyl's operational definition of (inertial) mass (Jammer, 2000; Weyl, 1949).

of the set, their also exists an element in the set which is their concatenation. This is the case if we indeed help ourselves to the abovementioned full sets of properties with cardinality $\aleph_1$, even though this may not be ontologically parsimonious. However, in the first-order case, it seems to be empirically false that for every pair of massive objects in the actual world there exists a further object in our actual world with a mass that is the sum of the masses of the pair of objects. One might respond by adopting a version of mereological universalism: any fusion of two massive objects (i.e. a whole that is the mereological sum of those two parts/objects) is a further object, with a mass that is the sum of the masses of the two initial objects.[12]

A fan of absolute masses also has another option besides committing to an ontology of universals, sprinkled across the world in such a way that, for instance, the universal mass of an electron is instantiated multiple times at different locations given that there are multiple equally massive electrons. Instead, all these sprinkled-out properties are gathered together, so to speak, into one whole: a property space,[13] or determinable space, or magnitude space, or, in Wolff's terminology, a quality space – for the case of mass, Arntzenius and Dorr simply call it mass space (Arntzenius, 2012, chapter 8). Material objects are not massive because they *instantiate* or *have* mass properties but in virtue of being *located* in mass space, just as they are located in regular space (Wolff, 2020, section 7.3.2).

Wolff (2020), who advocates a version of this ontology, believes it to be a third option besides the absolutist's absolute masses and the comparativist's mass ratios. Strictly speaking she is right, in that this is not an ontology of *monadic* things. However, it is clearly in the absolutist spirit and far removed from comparativism. There are two reasons why it is monadic in a loose sense. First, it is single particles that are located in that space; it would not make sense to talk of a pair of particles having a single location in mass space. Second, mass space can have sufficient metaphysical structure to support active Leibniz scalings: to acknowledge the metaphysical difference that absolutists believe is needed to account for an empirical difference. Moreover, Arntzenius and Dorr (Arntzenius, 2012, p.229), whose mass space proposal inspired Wolff's account, seem to think that the mass space option compared to the universals option is a mere choice of terminology upon which nothing substantive hinges. That being said, the former seems to have an advantage over the latter. It explains the empirical fact that (pointlike, classical) objects only have a single, unique mass magnitude (if they indeed have magnitudes at all) (Wolff, 2020,

[12] I would like to thank Mahmoud Jalloh for providing me with this potential response.

[13] Funkhouser (2006) advocates a property space view.

p. 121; Mundy, 1987, p.40) in virtue of the classical interpretation of the *location* of a pointlike object in space – and now thus also in mass space – as being unique.[14] This ontology may also seem to come with a disadvantage: all electrons are co-located in mass space. However, to the extent that that may seem counterintuitive, that is because in regular space the gravitational and Coulomb laws would blow up if particles are at zero distance from one another. There is however no such dynamical problem within mass space itself.

In sum, two main options for an ontology of absolute masses are in terms of (universal) properties or mass spaces, and two main options for an ontology of scale-invariant mass relations are in terms of (universal) properties or by applying ordering and concatenation relations directly to objects (whether those are material objects or spacetime points). For each of these, one further decision point remains: whether to add so-called *primitive/intrinsic/non-qualitative identities* (Figure 6). Two absolute magnitudes are not absolutely discernible (cf. the representational indeterminacy/kinematic irrelevance of absolute masses in Section 1): that is, there is no logical formula with one free variable which is true of one of the magnitudes but not the other (even though they are of course relatively discernible, i.e. there exists a formula with two free variables, e.g. $m_i > m_j$, which applies in one order only) (Jacobs, 2023b; Ladyman, 2007; Quine, 1960). After all, the representation of the absolute masses by numbers was only unique up to a constant. In the case of mass ratios only the identity element is absolutely discernible from any other element (as only the identity element is such that concatenating it with itself returns itself) and any other two elements are not absolutely discernible. From one numerical representation of the mass ratios one can always obtain an equally faithful representation by, for example, squaring all the numbers (which leaves unchanged only the number 1, representing the identity element). Consider the case of absolute masses. Consider furthermore the interpretation of laws of nature as *governing* what happens to objects with absolute masses (but see Section 6.2 for an alternative interpretation), in the sense that one starts with the fundamental ontology at an initial time – the initial state – and the laws make the future happen by evolving that initial state state forward. If the realist about absolute masses is right that they are dynamically relevant – that an active Leibniz scaling of the initial masses leads to an observationally distinguishable world – then the metaphysics must be rich enough to indeed distinguish the two initial states of these two possible worlds. As it stands, with absolute masses not being absolutely discernible, that is not the case. One would need to add further primitive, intrinsic

[14] Wolff (2020, p.121) points out that this single-value principle is also explained on (other) first-order views where the ordering relations apply directly to objects.

identities to these absolute masses – one may think of these as transworld names that allow one to identify the masses across possible worlds – in order for the governing laws to know which mass they are 'latching onto'; that is, given that in this world this mass accelerates this much, in the other world it is *that* mass and not *that other* mass that accelerates by the same amount as in the first world. More on this in Section 6.3.

The standard nomenclature for intrinsic identities attached to objects is haecceities, and when they are attached to properties they are called quiddities. Ontologies without such primitive identities are called anti-haecceitistic or anti-quidditistic or sophisticated (Dewar, 2019). For instance, Jacobs (2023b) is a sophisticated absolutist, and Wolff (2020), who advocates a mass space ontology but without primitive identities of the points in that space, refers to her view as sophisticated substantivalism (in the sense of sophisticated mass space substantivalism).

4 Intermezzo: Three Characterisations of Dynamical Relevance

We return to the *dynamical (ir)relevance* of absolute masses, as the literature shifts back and forth between at least three different ways of interpreting this notion: the symmetry approach, the undetectability approach, and the modal (or possibility checking) approach. Ideally these approaches would all be equivalent and thus interchangeable, but whether they indeed are turns out to be a subtle matter.

When dynamical (ir)relevance is understood in terms of symmetry (Dewar, 2020; Jacobs, 2023b; Jalloh, 2022; Wolff, 2020), the issue boils down to the anti-realists claiming that active Leibniz mass scalings are symmetries of Newtonian gravity and most realists denying this. Anti-realists sometimes combine this with an Occamist norm that makes a direct symmetry-to-unreality inference without even having to invoke parsimony considerations, for instance Saunders' (2007) invariance principle: "only [magnitudes] invariant under exact symmetries are real".[15] Unfortunately the notion of symmetry is a contentious topic within the philosophy of physics literature – see, for example, Dasgupta (2016) and references therein, and see, for example, Wallace (2022) for a response to some of the problems raised by Dasgupta. On some definitions of a symmetry transformations, it is not necessarily the case that they map empirically indistinguishable states of affairs to one another, which would

[15] See Møller-Nielsen (2017) and Martens and Read (2021) for a critique of these so-called interpretationalist symmetry-to-unreality inferences.

make this notion of symmetry useless for our purposes (or at least the purposes of the anti-realist). If we could have an independent grip on the notion of symmetry which would then allow us to go out in the world and *discover* that models related by such a symmetry are empirically equivalent (contra Read and Møller-Nielsen (2020)), this would be perfect. However, if symmetries are simply *defined* as any transformation between initial states that generates an empirically indistinguishable model then, even though one is perfectly allowed to use the term 'symmetry', the notion of symmetry is merely a name that does not provide any help over and beyond just talking about dynamical (ir)relevance and empirical indistinguishability directly. Anyway, in this case the symmetry approach is equivalent (by definition) to the modal approach that we will go on to discuss in this section.

Dasgupta (2013) favours the undetectability approach: framing the issue in terms of whether absolute masses are detectable or not (see Section 2.2). Here we need to distinguish between *detecting-which* (absolute mass *this* specific object has) and *detecting-that* (varying absolute masses uniformly leads to detectable differences). We do not even have to turn to the dynamics to figure out that one could never detect *which* specific quidditistic absolute mass *this* or *that* specific object has. This is because of (sophisticated) absolute masses being only relatively discernible (Section 3). However, this concerns subquestion 1 (kinematic relevance) of the Ozma mass problem (Section 1), not dynamic relevance. Whether the kinematic irrelevance is an epistemic vice/problem or not is a subtle issue, discussed in the literature under the label 'inexpressible ignorance' (see Maudlin (1993), Dasgupta (2015), Martens (2021, section 9), Martens (2017a, section 2.9) and Jacobs (2021, section 6.6)). However, even if it is, this does not save the anti-realist from having to answer the much more essential detectable-that question – subquestion 2a of the Ozma mass problem – of whether an active Leibniz mass scaling leads to detectably different possible worlds. This understanding of detectability then reduces to the modal approach.

With the modal approach, the anti-realist is challenged to recover all the possible worlds (up to empirical distinguishability) that are possible according to Newtonian gravity with (realism about) absolute masses, for example both escape and coincidence two-particle worlds (see the completeness condition in Step 5 of Table 1). In other words, the anti-realist is challenged to prove empirical equivalence with realism – to show that it is not empirically worse off. Thus, by ranging over all possible anti-realist initial states (i.e. without absolute masses) and evolving those forwards with an anti-realist-friendly form of the laws (Section 6), one should generate all the empirically distinct evolutions that the realist generates: for each possible world according to the realist, the anti-realist should generate – allowing this as being physically possible – an empirically equivalent possible world.

When phrased like this, it becomes clear that the most direct way for the anti-realist to push back against the realist is as follows. If we demand full empirical equivalence with realism about absolute masses, the deck seems stacked against the anti-realist.[16] This begs the question: it already assumes that the realist is right, and that the anti-realist must live up to the realist results, in the sense of recovering their full modal range of dynamical possibilities. All that can really be demanded is empirical adequacy, not full empirical equivalence with a position that the anti-realists disagree with. It may simply be the case that realism and anti-realism have different modal consequences; that is, they imply different sets of counterfactuals.

Compare this to the seemingly analogous case of Newton's bucket experiment. Suspend a bucket of water from the ceiling with a string. Rotate the bucket many times until the string is tightly wound. Wait a bit, holding the bucket still. The water will be flat, and at rest with respect to the bucket. Now let go of the bucket. The bucket will start twirling (with respect to the ceiling). Eventually friction will ensure that the water rotates along with the bucket – it returns to the state of being at rest with respect to the bucket – but this time the water will not be flat but concave. Substantivalists, that is those who believe that space exists independently from matter (i.e. the analog of realism, or strictly speaking absolutism, about masses), will explain the difference in observable effects – flat vs. concave water – in terms of the difference in angular momentum of the bucket-plus-water system – before and after letting go of the bucket – with respect to absolute space. Relationalists, that is those who deny substantivalism (i.e. the analog of comparativism), have no choice but to acknowledge the two observationally distinct situations – anyone could do the bucket experiment at home – but can find an explanation in terms of the different relative angular momenta of the bucket-plus-water system with respect to the ceiling and earth. However, these relationalist tools would not be available if we consider a world that contains only the bucket and water.[17] Substantivalists then typically make the further claim that in such a lone-bucket world – the supposed analog of our two-particle scenario (Section 2.2) – one would still have empirically distinct possibilities – a lone bucket with flat water and a lone bucket with concave water – which can once again be explained in terms of

[16] I would like to thank Chris Timpson for pushing me on this response.

[17] We are here glossing over the fact that without a ceiling and gravitational pull by the earth the water would not stay in the bucket, so we would have to tweak the experiment, for example by considering a closed bucket and measuring the 3D distribution of the water within the bucket.

different global angular momenta of the bucket with respect to absolute space. However, the relationalist seems at a loss in this scenario.

The easiest move for the relationalist is to dissolve the problem by denying from the start that all these dynamical possibilities need to be accounted for (see the completeness condition of Step 5 of Table 1). According to the Machian response, if there is nothing else in the world with respect to which the bucket could be rotating, then the water will simply be flat: the zero global angular momentum solution in substantivalist terms. The physical possibility of a lone-concave-bucket world is denied. This would make relationalism empirically inequivalent to realism, but does not thereby necessitate that it is empirically inadequate. We do not and will never have access to sparse worlds with just a bucket or just two particles, and the empirical data that we do have seems to be no problem for a Machian relationalist. That is, (1) the prevalence of both concave- and flat-water buckets in our actual, messy world can be accounted for by their differing spatiotemporal relations with respect to, for instance, the ceiling and the earth, and (2) measurements of the global angular momentum of our universe are consistent with a zero value.

There is, however, one crucial disanalogy between the bucket scenario and our two-particle scenario. In the former, it is obvious what the *deterministic*, unique evolution of a lone-bucket world is: the one that does not have any global angular momentum (with respect to absolute space), as this notion is exactly what the relationalist does not acknowledge. In our scenario, there is no obvious, default, null evolution.[18] Removing absolute masses and keeping only mass relations does not tell us whether two-particle worlds will be of the escape or coincidence type. Arbitrarily picking that they will be escape (or coincidence) worlds will not do the trick. The burden of proof rests on the anti-realist: how does two-particle-world behaviour – or really the behaviour of any world with massive objects – follow from the initial state plus the dynamics (both understood in an anti-realist-friendly way)? The anti-realist needs to respond to the threat of *indeterminism*.[19]

5 Decision Point: Other Initial Determinables

Quite generally in science – natural, life, social – if one wants to investigate (the influence or meaning or importance of) one concept, one parameter, one variable, then one attempts to perform a controlled (thought) experiment where only this parameter is varied and *everything else* is kept constant. This is also the

[18] See Martens (2017a, section 4.4.2) for an extended discussion.

[19] See Dasgupta (2020), who argues that some such forms of indeterminism are in fact a virtue rather than a vice.

idea behind the active Leibniz mass scaling (Section 2.2): if we want to determine the relevance of absolute masses, if any, we need to change only them, and leave unchanged anything else that is *independent* from these masses. The problem is determining this independence. Whether G should be understood as independent or not will be discussed in Section 6.3. Here we focus on the other determinables that one chooses to include in the initial state. When discussing the two-particle challenge to the anti-realist in Section 2.2, an implicit choice was made in this regard: distances and velocities were considered to be independent from mass. Although this choice is fairly standard, one is *prima facie* free to postulate a different set of initial determinables. And that choice will matter. An easy example: varying the masses while keeping the initial distances and velocities unchanged will also change the momenta ($p = mv$). However, if the initial state is given in terms of masses, distances and momenta, so that one considers not the velocities but the momenta as independent from the masses – as is common in the Hamiltonian framework – an active Leibniz mass scaling would leave the momenta the same, thereby changing the velocities instead.

More to the point, even if the realist about absolute masses ends up being right that the following scaling transformation of the initial conditions would lead to an empirically distinct evolution as it does *not* leave the escape velocity inequality invariant,

$$\begin{array}{lll} m & \mapsto & \beta m \qquad \beta \in \mathbb{R}^+ \\ r & \mapsto & r \\ v & \mapsto & v, \end{array} \tag{2}$$

the following transformation does leave the escape velocity inequality invariant and thus leaves the empirical evolution unchanged:

$$\begin{array}{lll} m & \mapsto & \beta m \\ r & \mapsto & r \\ v & \mapsto & \sqrt{\beta} \cdot v. \end{array} \tag{3}$$

(On the symmetry approach, we would say that even though transformation 2 is not a symmetry of NG, transformations 3 and 4 are. See also Baker (manuscript, June 2013), Dewar (2020) and Jalloh (2022) for a more general discussion of joint scalings.) Written in this way, transformation (3) seems illegal: we are falsely concluding that masses are dynamically irrelevant because we cheat by compensating with a change in an independent determinable, namely the initial velocity.[20] However, if we choose as independent initial variables the masses,

[20] As argued in Martens (2017a, section 4.2.1), Baker (manuscript, May 2013) cheats in the opposite way, by keeping the acceleration the same as if it were an independent variable even

distances, and a generalised notion of momentum $\wp \equiv \frac{v}{\sqrt{m}}$, the corresponding transformation *is* an appropriate active Leibniz mass scaling:

$$\begin{aligned} m &\mapsto \beta m \\ r &\mapsto r \\ \wp &\mapsto \wp. \end{aligned} \tag{4}$$

(Note that this enforces $v \mapsto \sqrt{\beta} \cdot v$.) When the *ceteris paribus* clause in the active Leibniz mass scaling is cashed out in this sense, absolute masses are not dynamically relevant. To what extent an alternative set of initial variables really vindicates anti-realism about absolute masses is discussed in Martens (2017a, section 4.2.2). The main point here is that the question of whether an active Leibniz mass scaling generates an observationally distinct possible world depends on which determinables one considers to be independent: that is, to constitute the initial state. This decision needs to be made in order for the issue of the dynamical relevance of absolute masses to be well-defined.

6 Decision Point: Laws and Constants of Nature

The dynamical relevance argument for realism depends, at least *prima facie*, on interpreting the laws as referring to initial absolute magnitudes which are then evolved forward in time, and especially and explicitly so in the crucial escape velocity inequality derived from the laws of Newtonian gravity. In this section we consider various alternative interpretations of laws of nature, and pay special attention to the role of dimensional constants of nature that appear in the laws. How one chooses to interpret the laws and constants of nature affects whether one does or does not need to postulate absolute magnitudes, and whether they require quiddities.

6.1 Absolute-Magnitude-Free Laws

A way of interpreting laws of nature in the spirit of anti-realism about absolute magnitudes is by understanding the laws as referring only to ratios ('of absolute magnitudes'). This interpretation is favoured by, for instance, Dasgupta (2013) and Roberts (2016).[21] When Newton's second law is written as $F = ma$ this may look as if the law relates three absolute magnitudes, but we are being misled. This notion is a shortcut for

$$\frac{F_1}{F_2} = \frac{M_1}{M_2}\frac{a_1}{a_2}, \tag{5}$$

though it is in fact dependent on the masses (if the distances and velocities are considered to be independent).

[21] It seems that Peacocke (2019, p.46) understands laws in this way, even though he also seems to talk about magnitude realism at times.

where the indices refer to one of two particles. A similar interpretation of the law of gravitation might look like this:

For any material things w, x, y and z in the same world,

(a) For any reals r_1, r_2 and r_3, if w has r_1 times as massive as y and x has r_2 times as massive as z, and if the distance between w and x is r_3 times as large as the distance between y and z, then the equal but opposite force between w and x is $\frac{r_1 \cdot r_2}{r_3^2}$ as strong as the equal but opposite force between y and z.

(b) For any real r_4, if the force between w and x is r_4 times as strong as between y and z, then there are reals r_5, r_6 and r_7 for which $\frac{r_5 \cdot r_6}{r_7^2} = r_4$, such that w has r_5 times as massive as y and x has r_6 times as massive as z and the distance between w and x is r_7 times as large as the distance between y and z.

Note that Newton's Constant of Gravitation 'falls out' – its strength is meaningless. Interestingly, Newton himself did not introduce the constant named after him, but talked in terms of proportionalities. As far as I am aware, König and Richarz are the first to explicitly mention the gravitational constant in print, in 1885![22]

A potential problem with these absolute-magnitude-free versions of the laws is that it is unclear that they are indeed still laws, rather than merely factive statements which can be checked to be true or false for a given possible world. That is, they are not in the form of differentiable equations, which would have supported solving initial value problems.[23] Whereas the realist accused the generic anti-realist of an initial state without absolute masses being indeterministic – allowing too many evolutions – these specific anti-realist laws seem to do even worse by providing no evolution whatsoever.

It may be retorted that such an understanding of laws of nature, in the tradition of Poincaré and Laplace, is too narrow. If laws manage to constrain the whole set of possible worlds – that is, to correctly pick out the set of possible worlds that should be dynamically allowed and rule out the worlds that should not be (see the modal approach in Section 4) – is that not all we can ask for?

This interpretation of the laws still faces two problems. First, these laws allow too many possible worlds; they allow worlds that we tend to consider dynamically impossible (a violation of the soundness condition in Step 5 of Table 1). If we start with a 'good' evolution, say an escape evolution, then multiplying the velocities or accelerations by any number still satisfies the laws

[22] I would like to thank Isobel Falconer for this information.

[23] As Mahmoud Jalloh has pointed in personal communication, this (as well as the oscillation problem that I also discuss in this section) raises the fascinating and paradoxical historical question – given that Newton himself talked in terms of proportionalities – of how Newton did manage in practice to do any physics, that is to (implicitly) solve initial value problems.

(if they are understood as we have discussed) as long as all velocities or accelerations are multiplied by the same number at a specific time, and thus produces another 'good' solution. Thus we could, for instance, multiply with a function that oscillates over time, such that the two particles first move away from each other and then approach each other and then move away again, etc. If this oscillatory pattern is symmetrical, the 'laws', understood as ratios of magnitudes at each time, are statements that are true of the oscillating system.

The second problem is that this approach seems to circumvent rather than solve the original problem. It still does not answer the question of whether the two-particle scenario described in section 2.2 results in escape or coincidence (see Section 4). The anti-realist may deny that two-particle evolutions need to be accounted for. Deriving the escape velocity inequality from these two interpretations of the laws results in an equation that can only be understood when referring to four or more particles.[24] Consider a world with one pair or equally massive particles (at a distance r and relative outward velocity of v) and another pair or equally massive particles (at a distance r and relative outward velocity of v) so far away from the first pair as to be effectively dynamically isolated. In that case, if the first pair, for instance, just about escapes each other, and the second pair stands in a 2:1 mass ratio to the first pair, then the absolute-magnitude-free escape velocity inequality has no problem predicting that the second pair will be a collision pair.

However, this only disguises the original problem. How can we predict the behaviour of the *first* pair? The issue was never how to predict the behaviour of subsequent subsystems with a determinate mass relation to a subsystem of which the behaviour is known. After all, from the realist perspective, knowing one absolute mass plus all mass ratios in the world fixes all the absolute masses of the other massive objects. Thus, if the behaviour of the first subsystem is known – which equates to the absolute mass(es) being fixed – then that of all other subsystems follows from the inter-system mass ratios. The thing at issue here was only ever a single degree of freedom: the absolute mass scale. The anti-realist wielding the absolute-magnitude-free interpretation of the laws will have to provide solutions to these two problems.

6.2 Humean Laws

It is possible to accept the realist's argument, but nevertheless to insist that our fundamental ontology – the things that exist independently from each other when building a possible world from scratch – needs only mass ratios. In other

[24] It could be argued that three particles suffice, as this already allows for two independent mass ratios, two independent velocity ratios, etc. Here we'll focus on four particles for simplicity.

words, we combine realism with comparativism. This approach works best with Humean interpretations (also known as regularity interpretations) of laws of nature. Instead of understanding laws of nature in the 'governing' sense – one starts with the fundamental ontology at an initial time (the initial state) and the laws make the future happen, by evolving forward the initial state (see Section 3) – as was implicitly the case in Section 2.2, the Humean interpretation flips this initial value problem approach on its head. The fundamental ontology includes a 3+1 dimensional mosaic criss-crossed with, for instance, particle trajectories. One may add further structure to this basic mosaic. Traditionally, intrinsic properties of the particles, such as absolute masses and absolute electric charges, are sprinkled 'onto' the spatiotemporal mosaic. Laws are then nothing more than concise statements that summarise the behaviour of particles within this four-dimensional mosaic, in a way that simultaneously maximises the competing virtues of simplicity and informativeness (i.e. they must rule out neither too many nor too few possible worlds). Laws do not make the world happen (from an initial state) but are just bookkeeping devices. A regularity comparativist modifies this account by sprinkling only mass ratios onto the spatiotemporal mosaic, not underlying absolute masses (by which the mass ratios would then be fixed) (Martens, 2017b). Thus, from an 'escape mosaic', that is a four-dimensional mosaic with two particles of, say, equal mass which move further and further apart, not only the gravitational law is reconstructured but also, simultaneously, its constant G and the absolute masses of the two particles (or at least the combination Gm).[25] On the one hand this regularity comparativist approach vindicates realism, as it accepts the dynamical relevance argument, but on the other hand it also deflates it, since the absolute masses can be obtained 'for free' from a mass-ratio mosaic – the *fundamental* ontology requires (at most) mass ratios but no absolute masses.

It is because the original version of the regularity approach included the most rich conception of mass (i.e. absolute masses) that the subsequent question became: can we do with a 'weaker'[26] mass structure, that is, mass ratios? However, if one would construct mosaics from scratch, the first question to ask is whether we need to add any mass structure to the mosaic at all. We do not. The governing approach derives (relative) particle trajectories – that is, distance ratios over time – from an initial state and nomic structure (i.e. laws including any constants appearing in those laws). The regularity

[25] Note that in this case the absolute masses would still be monadic, but would be extrinsic rather than intrinsic, as they are derived from the complete mosaic.

[26] As noted in Section 7.2, it is not obvious that a fundamental ontology of mass ratios is indeed 'weaker' than an ontology of absolute masses, in the sense of being more ontologically parsimonious.

approach inverts this initial variable problem: it reconstructs the initial state and the nomic structure from distance ratios of particles over time. A purely spatiotemporal mosaic with particle trajectories will also allow one to reconstruct mass ratios. The approach works even better than intended. Thus, the regularity/Humean approach suggests eliminating mass from the *fundamental* ontology altogether – even though it is still a view that is *realist* about absolute masses, such that the non-fundamental ontology does include absolute masses.

As a side-note, we might ask whether mass could also be eliminated in the context of a governing-law understanding of the dynamics, in the sense that an initial state of purely spatiotemporal determinables – distance, velocity and acceleration – would suffice for a) a deterministic evolution, and b) recovering the same possible worlds as the original theory that does include mass.[27] In effect, masses would be replaced by accelerations (cf. footnote 20). The easiest way to see that the answer is negative is via examples such as those in Figure 7, which show that a single spatiotemporally-described initial state is compatible with different masses – not only the absolute masses but also the mass ratios – in a way that leads to empirically distinct evolutions. A more detailed explanation of this impossibility of eliminating mass is given in Martens (2018).

6.3 Co-scaling G

Laws of nature, when expressed in terms of mathematical equations, typically include constants of nature, most of which are dimensional. Newtonian gravity is no exception: it features Newton's gravitational constant G which has the dimensions $\frac{\text{length}^3}{\text{mass}\cdot\text{time}^2}$. How can we interpret and use equations that feature such constants? Some scholars, such as Roberts (2016) and Jalloh (2022), claim that the exclusive Leibniz mass scaling defined in Section 2.2 is not the appropriate transformation. When scaling masses by a factor α, one also needs to scale G by a factor $\frac{1}{\alpha}$ because its dimensions include mass, and specifically mass^{-1}. It is this co-scaling transformation, which Jacobs (2023b) calls an inclusive active Leibniz mass scaling, that is the appropriate transformation to be considered. This inclusive scaling *is* a symmetry of the escape velocity inequality, as $Gm \rightarrow \frac{G}{\alpha}\alpha m = Gm$. Scaling absolute masses makes no dynamical difference, as long as one co-scales G.

One may even – but does not have to – go one step further and claim that this shows that the proper (i.e. the most concise, least redundant) way of representing the laws of nature is in the absolute-magnitude-free way discussed

[27] See Perry (2015, forthcoming) and Dees (2018), who advocate different senses of reducing mass/quantitative structure.

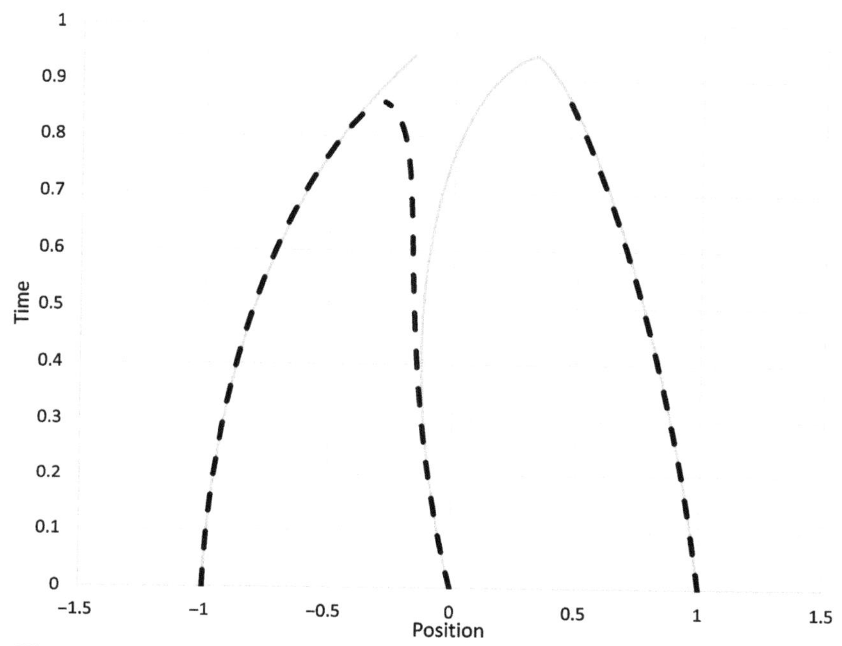

Figure 7 A numerical solution of the three-body problem in one dimension (dashed-line trajectories), superimposed on an alternative solution (solid-line trajectories). Each three-body problem has only been solved until the first collision, as the theory breaks down at that point. The initial states of each set of three particles are identical with respect to the spatiotemporal quantities ($G = 1, r_{12} = r_{23} = 1, v_1 = 0.1, v_2 = -0.6, v_3 = -0.3, a_1 = 1, a_2 = 1.5, a_3 = -0.625$), but they differ in terms of their masses ($m_1^d = 1, m_2^d = 0.375, m_3^d = 2.5; m_1^s = 0.1, m_2^s = 0.6, m_3^s = 1.6$). Subscripts refer to particle number.

in Section 6.1, where G drops out altogether. This reveals that co-scaling suffers the same potential problems as that approach. What is the evolution of a two-particle world, escape or coincidence – but perhaps this is just a brute fact depending on the strength of G? And shouldn't both evolutions be dynamically possible?

Let us return to the question of justifying the appropriateness of the inclusive over the exclusive mass scaling. One may think that the strength of G, as represented by a specific numerical value after fixing the units, is supposed to be fixed. Jalloh (2022) refers to this understanding of dimensional constants of nature as *constant necessitism*. An exclusive scaling would change the strength of gravitation and hence turn a solution of Newtonian gravity into something that is not a solution, or rather a solution of a similar but distinct theory with stronger or weaker gravitational dynamics (Martens, 2022). But perhaps this is too narrow-minded a view of what counts as a theory (Jacobs, 2023b), and we should allow G to vary within a single theory. Jalloh (2022) refers to this less restrictive understanding of dimensional constants of nature as *constant contingentism*.

The question remains, even if solutions are preserved under co-scaling of G, why the inclusive scaling would be not only a possible transformation but also the only correct way of scaling the absolute masses. As indicated in Section 5, it is quite generally the case in science that if one wants to investigate (the influence or meaning of) one concept, one parameter, then one attempts to perform a controlled (thought) experiment where only this parameter is varied and *everything else* is kept constant. Insisting that G must be scaled alongside the absolute masses seems to attach too much value to the representation of G in terms of a numerical value and a unit, as if it were yet another determinate physical mass-like magnitude that is predicated of objects out there in the world in the same way mass is. However, G is merely a way of characterising different choices of the strength of the gravitational law – this law being a function![28] Let us call this function $\mathcal{G}$. It maps initial variables to accelerations. For instance, for the standard choice of initial variables, $\mathcal{G} : M^2 \times R \rightarrow A$. This is illustrated in Figure 8.

In this figure we consider, for simplicity, two-particle scenarios with equal masses. The distance and initial velocity are kept constant across each frame of the figure and are omitted. The $\mathcal{G}$ of standard Newtonian gravity requires input and output variables with quiddities (Frame 8(a)); otherwise there is no way of comparing the behaviour of $\mathcal{G}$ across possible worlds: Frame 8(b) and Frame 8(c) would represent the same metaphysics since the absolute magnitudes are not absolutely discernible. $\mathcal{G}$ is a function, so it needs to 'know' *which* (mass) determinates need to be mapped onto *which* (acceleration) determinates – even though we as humans will not be able to detect *which* specific quidditistic mass a specific object has (see Section 4). With quidditistic properties, the behaviour of $\mathcal{G}$ in various possible worlds, for example a world with equally massive particles each with quidditistic mass Alice vs a world with equally massive particles each with quidditistic mass Bob, can be depicted in a single picture – Frame 9(a). When representing the absolute magnitudes numerically, one is allowed to choose whatever unit convention one prefers. Fixing these units will then also fix the numerical value G as it appears in the equation $F_g = \frac{GM_1M_2}{r^2}$. A change of units, that is the passive scaling depicted in Frame 9(b), corresponds to a mere change in representational convention. This will change the numerical value G, but it does not change the underlying

[28] Jacobs (2023b) provides an alternative view, where G is not part of the dynamics, but of the kinematics.

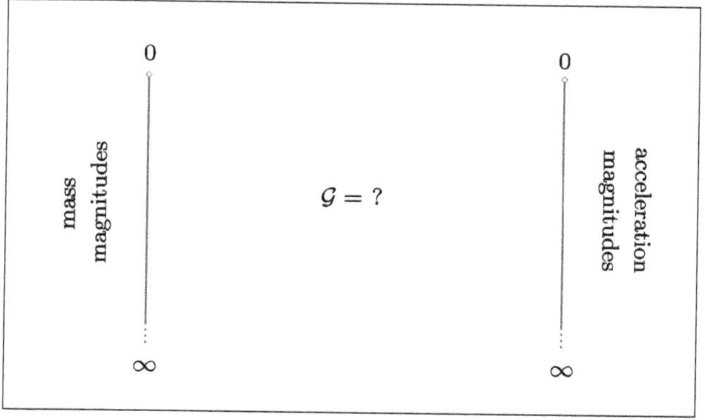

(a) Without quiddities – primitive, non-qualitative identities of absolute magnitudes across possible worlds – it is unclear which initial masses are to be mapped, by $\mathcal{G}$, to which accelerations.

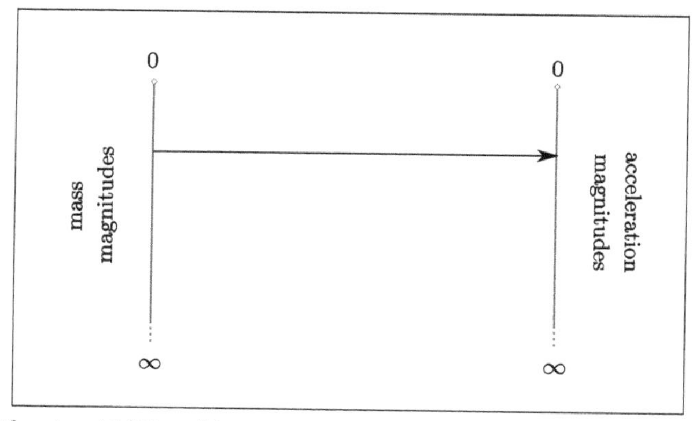

(b) Without quiddities this mapping could represent the same metaphysics as the next frame.

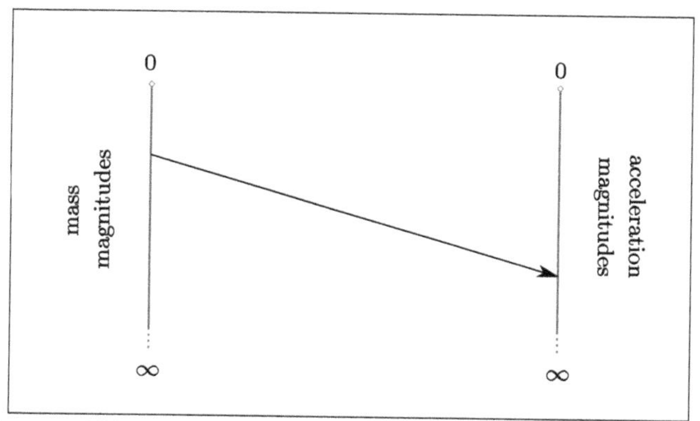

(c) Without quiddities this mapping could represent the sam metaphysics as the previous frame.

Figure 8 The gravitational law as a function from initial magnitudes to accelerations, $\mathcal{G} : M^2 \times R \rightarrow A$. In these figures, the distance and initial velocity are kept constant across each frame of the figure and are omitted. It is assumed for simplicity that both particles have the same mass.

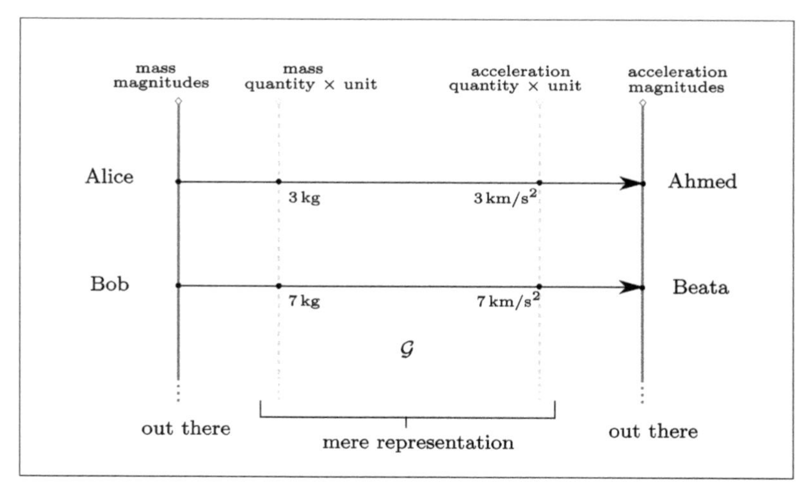

(a) Quidditistic masses and accelerations, and an arbitrary choice of units to numerically represent these (quidditistic) magnitudes.

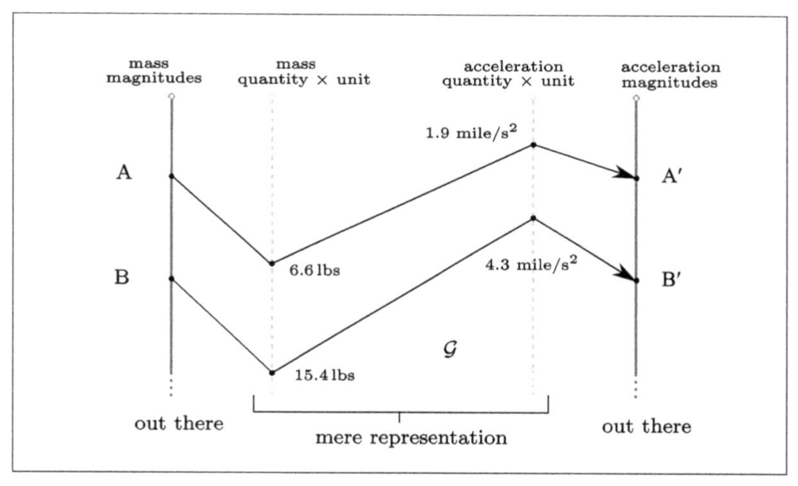

(b) Change of units compared to Frame (a), which constitutes merely a different representation of the same underlying function $\mathcal{G}$.

Figure 9 With primitive, non-qualitative identities, for example names such as Alice and Bob, the mapping $\mathcal{G}$ is well-defined (across possible worlds), and we can depict $\mathcal{G}$'s behaviour in two-Alice and in two-Bob worlds in a single picture. Fixing arbitrary units for all the determinables will also fix the value of the symbol G in those units. This will change if different units are chosen, but that will not change the underlying function $\mathcal{G}$. Hence, in subsequent figures we will omit these conventional numerical representations.

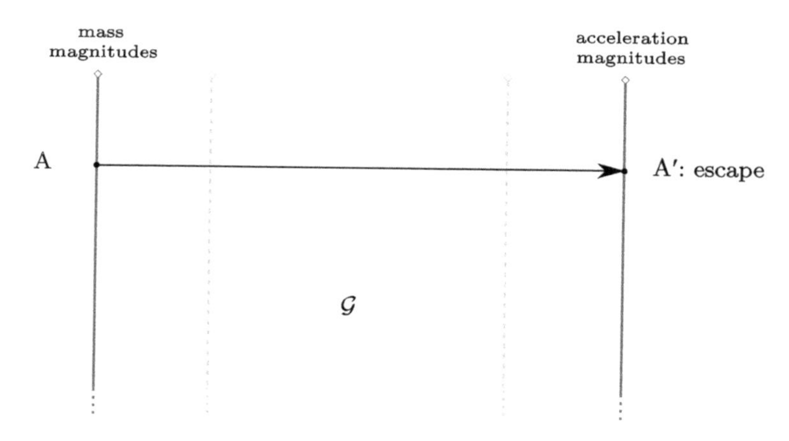

(a) Scenario 1: two-particle world with two Alice masses (cf. figure 4(b))

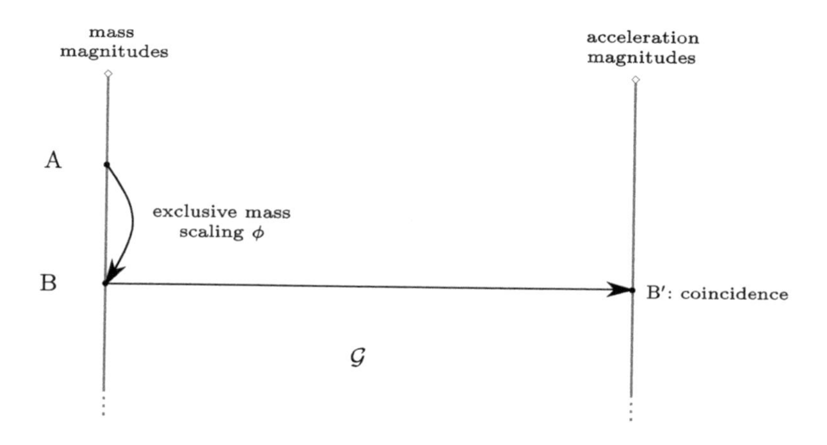

(b) Scenario 2: two-particle world with two Bob masses (cf. figure 4(a); obtained via an exclusive, active Leibniz mass scaling of scenario 1

Figure 10 The two-particle argument from Section 2.2. The laws are the 'standard' laws of Newtonian gravity, $\mathcal{G}$ (cf. Figure 9).

$\mathcal{G}$, which is a function from physical magnitudes to physical magnitudes 'out there' in the world, not as part of some numerical model/representation.

Consider first scenario 1 (Frame 10(a)), a possible world with two particles each with mass Alice, which is so small that under $\mathcal{G}$'s gravitational pull it gets mapped to an Ahmed (= A′) acceleration (see Frame 9(a)), corresponding (just about) to an escape evolution. Consider now scenario 2 (Frame 10(b)), which differs from scenario 1 by an exclusive scaling ϕ of all the masses, such that we arrive at a two-particle scenario each with the larger mass Bob. Since this is an exclusive scaling, $\mathcal{G}$ remains the same, which means that the Bob masses are mapped to Beata accelerations (Frame 9(a)), leading to a coincidence evolution. To repeat, this is empirically distinct from the escape evolution in scenario 1. Scenario 3 (Frame 11(b)) depicts an inclusive scaling: initial Alice masses are mapped to initial Bob masses, but – if we want to ensure that the accelerations do not change – we (must) also consider a distinct function, 'co-scaled $\mathcal{G}$', given by $\mathcal{G}' = \mathcal{G}^{-1}$ and depicted in Frame 11(a), which maps the Bob masses to Ahmed accelerations. Scenario 3 is then empirically indistinguishable from scenario 1.

Even if the inclusive scaling in scenario 3 is coherent, why would the exclusive scaling in scenario 2 not be the default scaling? Only if one attaches too much weight – pun intended – to the representation of $\mathcal{G}$ (or $\mathcal{G}'$) in terms of G, a number and a unit including a mass dimension, may one be led to believe that it would be inconceivable, incoherent or irrelevant to apply ϕ but not also $\mathcal{G} \rightarrow \mathcal{G}'$. $\mathcal{G}$ is a function that maps absolute physical magnitudes 'out there in the world' directly to accelerations, also 'out there in the world'. When trying to determine whether uniformly changing masses makes a difference, one just uniformly changes the input of the function, leaving the function alone. And even if we were to just stipulate, for whatever reason, that we must combine ϕ with $\mathcal{G} \rightarrow \mathcal{G}'$ and distance and velocity scalings with an analogous $\mathcal{G}''$, this would once again have the surprising consequence that if scenario 1 (escape) is possible then a coinciding two-particle world is dynamically impossible (or vice versa) (a violation of the completeness condition in Step 5 in Table 1)!

In sum, an anti-realist advocating inclusive scalings would have to justify why determining the dynamical relevance of masses *necessitates* changing the laws/ functions along with the scaling of the masses which are the input of those laws/ functions.

6.4 Machianism

The approach in the previous section attempted to reinterpret mass scaling (in the context of the standard equations of Newtonian gravity). A related, 'Machian' approach – with the same aim of rendering mass scalings as a symmetry of the set of laws – would be to consider distinct equations/laws from the start while interpreting their scaling in the traditional (exclusive) way (Martens, 2022). Mach tried to define inertial motion in terms of the bulk motion of matter in the universe (in an attempt to respond to Newton's bucket experiment – see Section 4 – which is conceptually very similar to the two-particle argument

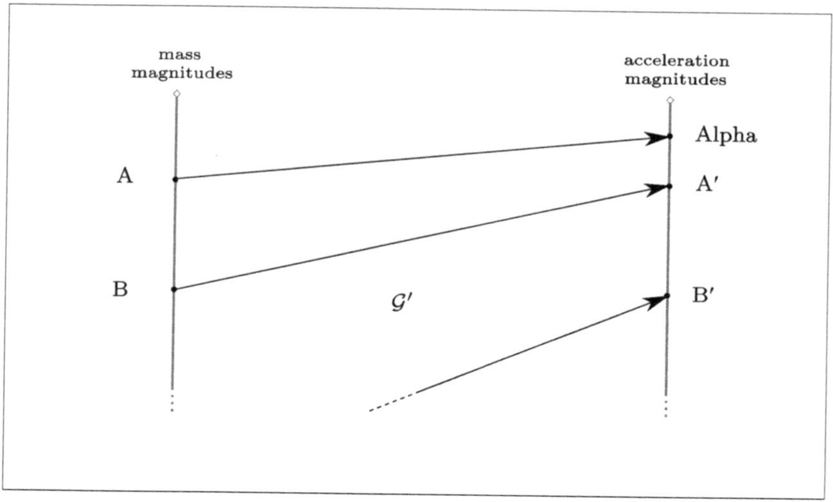

(a) The law of nature $\mathcal{G}'$ – a function – differs from the law of nature $\mathcal{G}$, as the latter is a different function. That is, $\mathcal{G}$ and $\mathcal{G}'$ map the mass A(lice) – which we can identify across this frame and Frame 9(a) because the primitive label/name A(lice) is a primitive, transworld identity – to different accelerations, namely Ahmed (= A′) and Alpha, respectively.

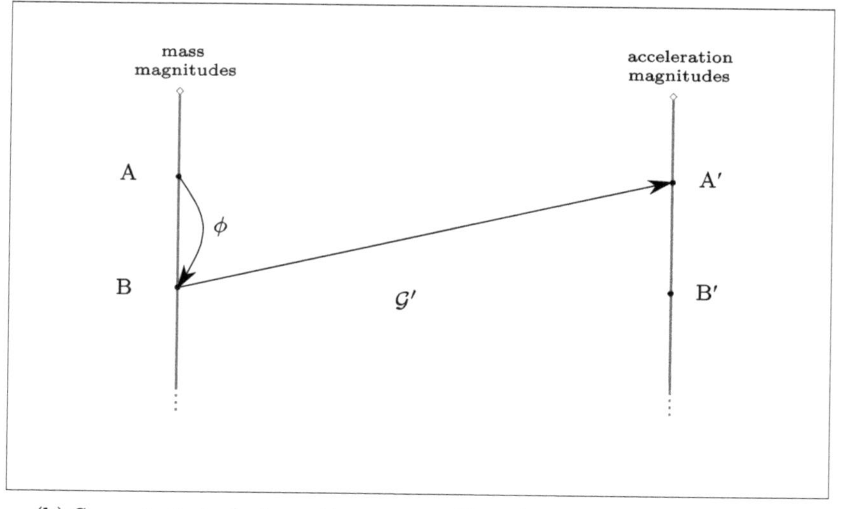

(b) Scenario 3: Inclusive active mass scaling of scenario 1 in Frame 10(a)

Figure 11 Inclusive active mass scaling, as opposed to the exclusive version in Figure 10(b). If one scales G along with the mass scaling (ϕ), and this G-scaling is intended to be an active scaling, this implies that one has moved from the function $\mathcal{G}$ to a distinct function, $\mathcal{G}'$, differing in the (quidditistic) acceleration that masses such as mass B are being mapped onto.

in Section 2.2). By analogy, we could replace G with $\tilde{G}/\sum_k M_k$, where $\tilde{G}$ is a constant (across possible worlds, space and time) with mass dimension zero and the sum is over all masses in the universe. The sum over absolute masses may seem an explicit commitment to realism, but in the relevant equations, $a = \tilde{G}\frac{M_j}{r^2 \sum_k M_k} = \tilde{G}\frac{1}{r^2 \sum_k \frac{M_k}{M_j}}$ and $v_{esc} = \sqrt{\frac{\tilde{G}}{r_0 \sum_k \frac{M_k}{M_j}}}$, only a sum over mass ratios appears.

Scalings of these syntactically distinct equations, representing the distinct – compared to $\mathcal{G}$ – function $\tilde{\mathcal{G}}$, are illustrated in Figure 12. (A picture like Frame 9(a) would be misleading, as we cannot superimpose different two-particle scenarios, because the behaviour of $\tilde{G}$ depends on all the massive particles in a given world.) Machian scenario 1 (Frame 12(a)) again represents a two-particle world with two Alice masses, and let us say that $\tilde{G}$ maps these to accelerations that are small enough for the particles to escape each other. Machian scenario 2 (Frame 12(b)) represents, as in Frame 10(b), an exclusive scaling of the masses compared to scenario 1, generating a two-particle world with Bob masses. However, this scaling is a symmetry of $\tilde{\mathcal{G}}$: the Bob masses get mapped to the same accelerations as in Machian scenario 1 because the new escape velocity inequality is a function of mass ratios rather than absolute masses, rendering both scenarios empirically indistinguishable. Note that we are now using a single function ($\tilde{\mathcal{G}}$) in both worlds, in contrast to Frame 10(a) versus Frame 11(b). $\tilde{G}$ having mass dimension zero eliminates any possible motivation for co-scaling $\tilde{G}$. The exclusive and inclusive (mass) scalings coincide – the choice of $\tilde{\mathcal{G}}$ forces the analog of Frame 11(b) (rather than Frame 10(b)) as the only conceivable option.

Now, if one had further modified the equations to include the appropriate sums of, for instance, distance and velocity so as to make uniform scalings of these magnitudes a symmetry as well, there would be no equally-massive-two-particle coincidence solution (if there is indeed an escape solution). Our old problem would return. However, such coincidence solutions do appear for $\tilde{\mathcal{G}}$ if for instance, one reduces the initial distance between the particles (Frame 12(c)). This does mean that one cannot save all possible evolutions if one is anti-realist about the absolute magnitudes of all the determinables involved (cf. the fully-absolute-magnitude-free laws in Section 6.1), but there is nothing about anti-realism that requires anti-realism across the board.

Interestingly, there is no empirical way of distinguishing between the standard and the Machian laws, as they agree on all dynamically possible worlds up to empirical equivalence: there is no independent way of determining which quidditistic r obtains in a world (i.e. there's no standard ruler that can be used

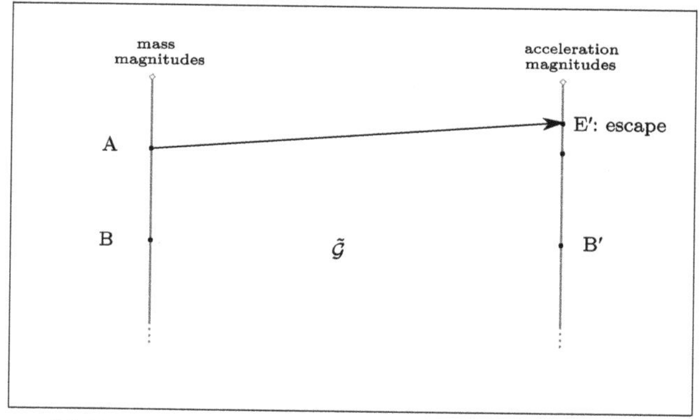

(a) Machian scenario 1: two-particle scenario with two Alice masses as in Frame 10(a) but with a different function: $\tilde{\mathcal{G}}$.

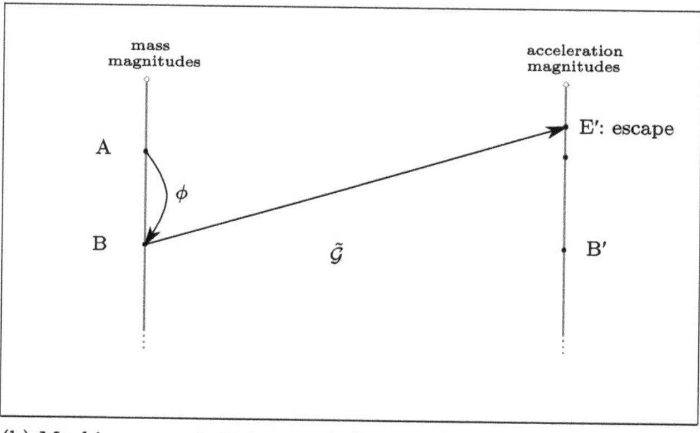

(b) Machian scenario 2: two-particle scenario with two Bob masses as in Frame 10(b) but with a different function: $\tilde{\mathcal{G}}$. Obtained via an exclusive, active Leibniz mass scaling of Machian scenario 1.

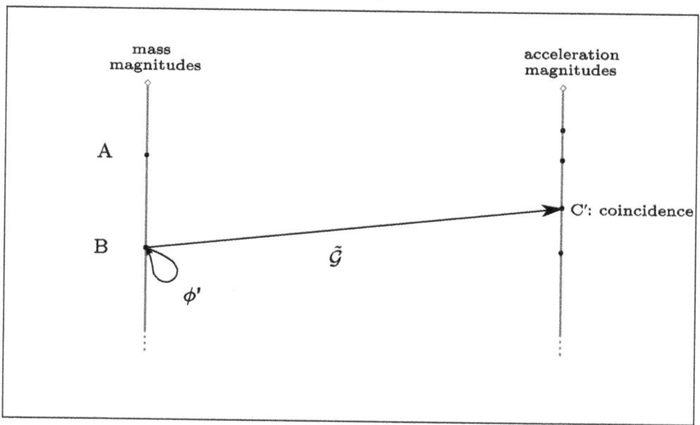

(c) Active Leibniz distance scaling/reduction ϕ', leaving the masses invariant

Figure 12 Machian Newtonian gravity, $\tilde{\mathcal{G}}$, a function that is distinct from the function $\mathcal{G}$ in Frame 9(a)

across possible worlds). Hence, there is no empirical reason that could forbid the anti-realist from favouring the Machian law over the standard law. Given this empirical underdetermination of the laws of Newtonian gravity, should we be agnostics about both the form of the laws and the existence of absolute masses at the same time? Or is there some Occamist norm that favours one package (Machian laws without absolute masses) over the other (standard laws with absolute masses)? Or could there be reason to insist on absolute masses (but perhaps without quiddities) even if one favoured the Machian form of the law? These questions and others will be the topic of Section 7.

6.5 Summary

Whether absolute masses are or are not dynamically relevant, which is the main point of contention between realists and anti-realists, depends on the laws of nature: their syntactic form (i.e. the equations), how one interprets that form, and how one interprets constants of nature that feature in the laws (see Steps 2 and 3 in Table 1).

The realist in Section 2.2 assumed the 'standard' form of the laws, $F_g = \frac{GM_1M_2}{r^2}$ and $F = Ma$, and interpreted those as representing (governing) functions from mass and distance magnitudes to acceleration magnitudes, with that function remaining invariant under an active Leibniz mass scaling: the scaling is exclusive, leaving G unchanged. Combined with, for instance, the 'standard' choice of further initial determinables, r and v, this interpretation of the laws renders absolute masses dynamically relevant.

Section 6.1 considered an anti-realist, alternative interpretation of the standard form of the laws, as referring to magnitude ratios only, not to (underlying) absolute magnitudes. For this view to stand a chance, one needs to solve the problem of too many possible worlds being allowed by this interpretation of the laws, and the problem of avoiding indeterministic evolutions of the anti-realist initial state.

Humeans disagree with the interpretation of laws of nature as governing the world, where starting from an initial (three-dimensional) state the laws govern/determine what happens from that moment onwards, generating the four-dimensional state of affairs. Instead, Humeans start with a four-dimensional mosaic and reverse-engineer the laws. The anti-realist modifies this framework by removing absolute masses from the Humean mosaic. The problem they would need to overcome is that this only shows that absolute masses are non-fundamental, but not that they are not real.

The fourth view sticks to the standard syntactic form of the laws, but claims that the appropriate mass scaling is not the exclusive version but the inclusive version: G needs to be scaled along with the mass magnitudes in such a way that overall no detectable change arises. Under such an inclusive scaling, the absolute masses are dynamically irrelevant. The main task for this view is to justify why the inclusive scaling is correct and the exclusive scaling is incorrect.

The final, Machian view claims that the way to achieve the intention of the previous view is to move to a syntactically distinct form of the laws, by replacing G in such a way that one is left with a constant of nature that has mass dimension zero, and an escape velocity inequality that depends on mass ratios and not on absolute masses. Since the mass dimension of this new constant is zero, the debate over inclusive versus exclusive scalings dissolves.

It is up to you, dear reader, to pick one of these five views, en route to determining the appropriate ontology of theories with dimensional determinables.

7 Decision Point: Non-dynamical Considerations

7.1 Explanatory Relevance

We now turn to Ozma mass subquestion 2b (Section 1), that is, the explanatory relevance premises in the main arguments in Section 2.1 (see also Figure 2): are there other, non-dynamical – or, more precisely, not as directly related to the dynamics as the two-particle scenario – empirical facts that require postulating absolute masses to be fully explained? In our universe, mass ratios turn out to be transitive. Whenever massive object 1 and massive object 2 stand in a numerical mass ratio M_{12} and massive object 2 and massive object 3 stand in a numerical mass ratio M_{23}, then the mass ratio between object 1 and object 3 is given, numerically, by:

$$M_{13} = M_{12} \cdot M_{23}. \tag{6}$$

Roberts (2016) calls this the ratio multiplication principle and traces it back to Russell (1903) and Armstrong (1988). See also Jacobs (2019, section 4.2) and Jacobs (2023a, section 4.4).[29] One may also call it the conspiracy of mass relations (Martens, 2022; Martens, 2017a, section 5.2.2), since from the perspective of the comparativist, for whom mass relations are (relatively) fundamental (i.e. they exist independently from absolute masses), mass relations should be independently specifiable – independently from any absolute masses *and* independently from one another. It is exactly this latter independence which M_{13} does not have in our universe. For comparativists, this transitivity of mass relations is a brute, unexplained fact. A surprise. A mystery. A conspiracy.

[29] Jacobs (2023a, section 4.5) discusses an interesting similar conspiracy in the context of electrodynamics, namely the composite loop multiplication feature satisfied by holonomies.

If, on the other hand, mass ratios obtain in virtue of absolute masses, as the absolutist claims, then their transitivity *is* explained. In fact, it is explained in the strongest sense possible, as it becomes a mathematical necessity:

$$M_{13} = \frac{M_1}{M_3} = \frac{M_1}{M_2} \cdot \frac{M_2}{M_3} \tag{7}$$

From this perspective there is no surprise, no conspiracy, whatsoever.

Thus, even if one thinks that absolute masses are not dynamically relevant, for example if one adopts Machianism (Section 6.4), one may still be persuaded to commit to absolute masses purely for their role in explaining the observed transitivity of mass relations. In that case sophisticated absolute masses would suffice, as quiddities are not needed to support any dynamical relevance of absolute masses.

7.2 Ontological Parsimony

You have now been provided with all the decision points for deciding between realism and agnosticism (Section 2.1 and Figure 2). That is, if one is persuaded by the explanatory relevance of absolute masses (Section 7.1) or by the dynamical relevance of absolute masses (Table 1) – after having chosen other initial determinables (Section 5) and a formulation and interpretation of the dynamical laws and Newton's constant (Section 6) – or by both, one then decides between an absolute mass space or absolute mass properties, as well as whether either of these includes quiddities (Section 3 and Figure 6). Whether one decides for or against quiddities depends strongly on the earlier decisions. Those who are not convinced of any relevance of absolute masses and therefore are not convinced of their reality are (tentative) agnostics about absolute masses. If you are a tentative agnostic who wants to make the stronger case for anti-realism (Section 2.1) – leading to a choice between first- and second-order mass relations (Section 3 and Figure 6) as fundamental ontology – then considerations of ontological parsimony come into play.[30]

It is sometimes taken for granted that anti-realism about absolute masses is more ontologically parsimonious than realism. The underlying intuition is presumably that anti-realists commit to – and hence pay the metaphysical price for – mass relations only, with realists committing to and hence paying for both absolute masses and mass relations. This is also how the main anti-realist argument in Section 2.1 is formulated. However, this is implausible. When a world is constructed, one does not have to add/pay for all the things that exist, only the things that are fundamental: that enjoy an independent existence. Once all

[30] This section draws heavily upon Martens (2022, section 7).

the fundamental things are added to a world, there is no freedom left, no choice left, no further price to pay: all non-fundamental things are automatically there. After all, them being non-fundamental *means* that their existence derives from – is determined by – the existence of the (more) fundamental things. Schaffer (2015) refers to this as replacing Occam's razor – do not multiply entities or properties without necessity – with Occam's laser – do not multiply *fundamental* entities or properties without necessity. An absolutist for whom facts about mass ratios follow from dividing the numbers used to represent absolute masses gets those facts for free. Similarly, a regularity comparativist does not include absolute masses in their fundamental mosaic but gets them for free (together with the laws, as a package deal) when concisely summarising that mosaic. This requires a revision of Section 2.1's argument:

P_1	**Dynamical irrelevance:** Absolute masses are dynamically irrelevant.
P_2	**Explanatory irrelevance:** Absolute masses are explanatorily irrelevant.
P_3	**Occam's laser:** All other things being equal (i.e. absolute masses are empirically – both dynamically and explanatorily – irrelevant), the more parsimonious *fundamental* ontology is the correct one.
P_4	**Ontological expensiveness:** A fundamental ontology of only mass ratios is more parsimonious than a fundamental ontology of only absolute masses.
C	Anti-realism about absolute masses.

Now that we are no longer comparing the parsimony of one ontology to that of a second ontology that is a proper subset of the first – in which case the second would have trivially been the most parsimonious – the 'ontological expensiveness' premise becomes much more difficult to evaluate. We need a measure of ontological parsimony, allowing us to compare two fundamental ontologies that are two disjoint sets (Martens, 2022, section 7). This requires various decisions, and it will be difficult to know how to make them.

Is it only quantitative parsimony that is relevant: minimising the cardinality of the fundamental ontology? Or is it only qualitative parsimony that is relevant: the number of types of things (entities, properties, etc.) in the fundamental ontology – or are some types more costly than others? Or are both quantitative and qualitative parsimony relevant; but then, how can we weigh their relative importance?

Quantitatively, for a world consisting of n particles, a fundamental ontology of absolute mass universals would comprise only n instantiations of such properties, whereas a fundamental ontology of mass ratio universals *prima facie* comprises $n(n-1)$ (or perhaps even n^2 if we include fundamental mass relations between a particle and itself) instantiations. One may try to improve upon the latter by not requiring a complete graph of relations between all n particles, but only a chain: the ith particle only instantiates one mass relation with the $i-1$th particle and a second one with the $i+1$th particle (with the first and last particle of the chain only being part of a single instantiation of a relation). This would require only $n-1$ instantiations of a mass relation, and would furthermore evade the conspiracy of fundamental mass relations discussed in the previous subsection. However, the first-order sympathiser would immediately point out that for these universal properties one is really committed to the Platonic existence of $\aleph_1$ properties in both cases. Furthermore, in the case of absolute masses, what happens if we adopt a mass space ontology instead? Does this count as one entity, or is this just $\aleph_1$ things in disguise?

Qualitatively, should all types of entities/properties be considered equally parsimonious, or not? Are entities on a par with properties? Are fundamental absolute mass universals on a par with fundamental mass relation universals? One may approach this latter question by comparing their structures. As indicated in Section 3, these are structurally quite similar, in that both form a total order and obey a concatenation rule. However, does the difference between monadic and dyadic properties matter for parsimony? And how costly is it to add primitive identities? How does a full mass space compare to absolute mass properties? Note also that Machian laws (Section 6.4) combined with anti-realism about absolute masses requires additional structure to support the sum of mass ratios over and above its multiplicative structure.[31] After all, otherwise the sum would differ between one numerical representation of the ratios and another one that consists of the squares of the first.

The discussion so far has considered measures of parsimony within a possible world. A final consideration concerns an inter-world parsimony measure: minimising the number of metaphysically distinct possible worlds per empirically distinct possible worlds. To paraphrase Occam: do not multiply possible worlds beyond necessity. *Prima facie*, anti-realism is more parsimonious than realism: to every set of mass ratios (as described by the initial state of a bunch of particles) there corresponds an infinite number of possible distributions of absolute masses, all differing by an active Leibniz mass scaling. However, if

[31] This sum is well-defined if absolute magnitudes do exist (Scott (1967), as cited by Peacocke (2019, p.45)).

one is an anti-realist about masses who distributes initial mass relations as a chain (as discussed in the preceding paragraphs) rather than as one relation between each pair of masses, there is a large number of possible ways to form that chain, leading to metaphysically distinct possible worlds without an empirical difference.

In sum, to go beyond agnosticism, the anti-realist about absolute masses would need to justify which of these parsimony considerations are to be included; to explain how they are weighed relatively to another, ideally all in a way that the realist can agree on; and to show that the result favours anti-realism.

8 Overview

Central features in our scientific theories, especially in physics, chemistry and the life sciences, are physical determinables with quantitative determinates, for example charge, length, mass and duration. Most of these are dimensional: their determinates are typically represented by a number (i.e. numerical quantity) and a (conventional) unit. What is the world like, to the extent that it is correctly described by theories with such physical determinables? A main point of contention is whether the correct (fundamental) ontology is one of absolute magnitudes underlying such a determinable, or magnitude relations that are represented by ratios (Figures 1 and 2).

First one has to determine whether a specific absolute magnitude in a specific theory is dynamically relevant (Section 2.2, Section 4 and Table 1). This requires deciding on the appropriate other determinables that constitute the initial state (Section 5), and then deciding on the form and interpretation of the laws of nature and any dimensional constants that feature in those laws (Section 6).

Second, one has to determine whether the specific absolute magnitude in the specific theory is explanatorily relevant, beyond being relevant in explaining the dynamics (Section 7.1).

If absolute magnitudes are relevant in at least one of these two senses, the final decision to be made is between an ontology of absolute magnitudes as universal properties or an ontology in the form of a magnitude space (Section 3 and Figure 6) (unless one combines Humeanism about the laws with comparativism). One's analysis of the laws of nature will play into whether either of these two ontologies requires additional quiddities or not.

If absolute magnitudes are not relevant in either of these two senses, one has to decide between agnosticism about absolute magnitudes, or attempting to make the further case for anti-realism about absolute magnitudes (Section 2.1).

This requires defining and justifying a relevant measure of ontological parsimony, and showing that this measure favours a (relatively) fundamental ontology of only magnitude relations and no absolute magnitudes (Section 7.2).

The final decision for the anti-realist is then between a second-order (relatively) fundamental ontology of mass relations as universal properties, or a first-order (relatively) fundamental ontology where the quantitative structures are instantiated directly by objects (Section 3). In the latter case, one has to determine whether those objects are material objects or spacetime points.

For the sake of concreteness, this Element has focused on the case study of mass within Newtonian gravity. However, the discussed here applies in the same way for other determinables and/or for other theories. For instance, one may consider an active Leibniz mass scaling in the two-particle scenario in Section 2.2; an active electric charge scaling in a similar scenario but governed by (an additional) Coulomb's force; scaling the masses[32] of a binary black hole system in general relativity, or some other famous examples:

- **Galileo's table:** Roll a ball across a tabletop. Once it leaves the table its path will be a parabola that eventually intersects with the floor. If instead we roll the ball with a different velocity or, equivalently (assuming that the length of the table remains the same), we scale the time, the shape of the parabola will differ – relative to the size of the table.
- **Poincaré's/Galileo's size doubling:** Would we notice anything if the size of everything in the universe changed overnight? Nehrlich discusses this thought experiment as follows:

> Perhaps the reason why Leibniz never used the doubling argument is that he read the first of Galileo's *Two New Sciences*. Galileo points out that doubling a thing in all its lengths (scaling it up) will quadruple its cross-section areas and multiply its volume and thus its mass by eight. The load-bearing factors of the object are essentially functions of area, but the loads they must bear for the thing to be stable are functions of its mass. So scaled-up models of strong small things, like ants, may be unstable, as producers of gigantism horror movies often find. The bigger they are the harder they stand – or fall. (Nerlich, 1991, p.172)

The result of the decision tree may well differ for other determinables and/or theories than mass in Newtonian gravity. The reader is invited to perform this analysis for their favourite determinable in their favourite theory.

Finally, let us return to the third Ozma mass subquestion: can we ensure that the aliens pick out a specific object with a mass that is the same as the mass

[32] The various notions of mass that are used within the context of general relativity are of course different from the mass determinable that we have been considering.

that we have labelled, say, '1 kg'? If anti-realism is true, then – as detailed in Section 1 – all that is meant by '1 kg' is that the object in question stands in an equal mass relationship to an arbitrarily picked object on Earth. No local experiment that the aliens could do will help them in determining which of their objects is 1 kg, but they would not be missing anything deep about the concept of mass; they would only lack knowledge of an arbitrary Earthly convention. If realism is true, then there is a restricted sense in which we can have them pick out a 1 kg object. That is, if we would first send them a rod and a clock, we can instruct them to use those to set up a two-particle scenario with exactly the correct initial r and v such that the escape mass is 1 kg, which would allow them to determine which objects have a mass that we have labelled as '1kg'. Thus, we would have reduced the Ozma mass problem to the Ozma length problem and the Ozma time problem. We have defined a dynamically privileged unit for mass, but only relative to a particular pair of rod and clock (Martens, 2021).

References

H. Alexander, editor. *The Leibniz-Clarke Correspondence*. Manchester: Manchester University Press, 1956/1717. Originally written by G. W. Leibniz and S. Clarke in 1715–16 and published by S. Clarke in 1717.

D. Armstrong. *A Theory of Universals: Volume 2*. Cambridge: Cambridge University Press, 1978.

D. Armstrong. Are quantities relations? A reply to Bigelow and Pargetter. *Philosophical Studies*, 54:305–16, 1988.

F. Arntzenius. *Space, Time, & Stuff.* Oxford: Oxford University Press, 2012.

D. J. Baker. Some consequences of physics for the comparative metaphysics of quantity. Manuscript, May 2013, later uploaded to http://philsci-archive.pitt.edu/12674/.

D. J. Baker. Comparativism with mixed relations. Manuscript, June 2013.

J. Bigelow and R. Pargetter. *Science and Necessity*. Cambridge: Cambridge University Press, 1990.

J. Bigelow, R. Pargetter and D. Armstrong. Quantities. *Philosophical Studies*, 54:287–304, 1988.

S. Dasgupta. Absolutism vs comparativism about quantity. In K. Bennett and D. W. Zimmerman, editors, *Oxford Studies in Metaphysics*, volume 8, pages 105–47. Oxford: Oxford University Press, 2013.

S. Dasgupta. Inexpressible ignorance. *Philosophical Review*, 124(4):441–80, 2015.

S. Dasgupta. Symmetry as an epistemic notion (twice over). *British Journal for the Philosophy of Science*, 67(3):837–78, 2016.

S. Dasgupta. How to be a relationalist. In K. Bennett and D. W. Zimmerman, editors, *Oxford Studies in Metaphysics*, volume 12, pages 113–163. Oxford: Oxford University Press, 2020. https://doi.org/10.1093/oso/9780192893314.003.0005.

M. Dees. Physical magnitudes. *Pacific Philosophical Quarterly*, 99(4):817–41, 2018. https://onlinelibrary.wiley.com/doi/abs/10.1111/papq.12223.

N. Dewar. Sophistication about symmetries. *British Journal for the Philosophy of Science*, 70(2):485–521, 2019. https://doi.org/10.1093/bjps/axx021.

N. Dewar. On absolute units. *British Journal for the Philosophy of Science*, 2020. DOI: 10.1086/715236.

M. Eddon. Quantitative properties. *Philosophy Compass*, 8(7):633–45, 2013.

B. Ellis. *Basic Concepts of Measurement*. Cambridge: Cambridge University Press, 1966.

H. H. Field. *Science without Numbers: A Defence of Nominalism*. Oxford: Basil Blackwell, 1980.

E. Funkhouser. The determinable-determinate relation. *Noûs*, 40(3):548–69, 2006. ISSN 00294624, 14680068. www.jstor.org/stable/4093996.

M. Gardner. *The New Ambidextrous Universe: Symmetry and Asymmetry from Mirror Reflections to Superstrings*. Mineola, NY: Dover Publications, Inc, 2005 (third edition), 1964/1990. Revised and retitled in 1990.

C. Jacobs. Gauge and explanation: Can gauge-dependent quantities be explanatory? Master's thesis, University of Oxford, 2019.

C. Jacobs. Symmetries as a Guide to the Structure of Physical Quantities. PhD thesis, University of Oxford, 2021.

C. Jacobs. Comparativist theories or conspiracy theories? The no miracles argument against comparativism. PhilSci Archive, Philosophy of Science Association, University of Pittsburgh, 2023a. http://philsci-archive.pitt.edu/21948/1/cosmic%20conspiracies.pdf.

C. Jacobs. The nature of a constant of nature: The case of *G*. *Philosophy of Science*, 90(4):797–816, 2023b. https://doi.org/10.1017/psa.2022.96.

M. Jalloh. The Π-theorem as a guide to quantity symmetries and the argument against absolutism. PhilSci Archive, Philosophy of Science Association, University of Pittsburgh, 2022. http://philsci-archive.pitt.edu/20743/.

M. Jammer. *Concepts of Mass in Contemporary Physics and Philosophy*. Princeton, NJ: Princeton University Press, 2000.

A. König and F. Richarz. Eine neue Methode zur Bestimmung der Gravitationsconstante. *Annalen der Physik*, 260(4):664–68, 1885. https://onlinelibrary.wiley.com/doi/abs/10.1002/andp.18852600409.

D. Krantz, R. Luce, P. Suppes and A. Tversky. *Foundations of Measurement, Volume 1*. New York: Academic Press, 1971.

J. Ladyman. On the identity and diversity of objects in a structure. *Proceedings of the Aristotelian Society, Supplementary Volumes*, 81:23–43, 2007. ISSN 03097013, 14678349. www.jstor.org/stable/20619100.

D. Lewis. *Philosophical Papers*, volume 2. Oxford: Oxford University Press, 1986.

N. C. M. Martens. Against Comparativism about Mass in Newtonian Gravity: A Case Study in the Metaphysics of Scale. PhD thesis, 2017a. https://ora.ox.ac.uk/objects/uuid:3f98e412-2cf7-4810-8a3a-0041f9c1c5df.

N. C. M. Martens. Regularity comparativism about mass in Newtonian gravity. *Philosophy of Science*, 84(5):1226–38, 2017b.

N. C. M. Martens. Against Laplacian reduction of Newtonian mass to spatiotemporal quantities. *Foundations of Physics*, 48:591–609, 2018. https://doi.org/10.1007/s10701-018-0149-0.

N. C. M. Martens. The (un)detectability of absolute Newtonian masses. *Synthese*, 198:2511–50, 2021. https://doi.org/10.1007/s11229-019-02229-2.

N. C. M. Martens. Machian comparativism about mass. *British Journal for the Philosophy of Science*, 73(2):325–49, 2022.

N. C. M. Martens and J. Read. Sophistry about symmetries? *Synthese*, 199:315–44, 2021.

T. Maudlin. Buckets of water and waves of space: Why spacetime is probably a substance. *Philosophy of Science*, 60:183–203, 1993.

T. Møller-Nielsen. Invariance, interpretation, and motivation. *Philosophy of Science*, 84(5):1253–64, 2017. https://doi.org/10.1086/694087.

B. Mundy. The metaphysics of quantity. *Philosophical Studies*, 51(1):29–54, 1987.

E. Nagel. Measurement. *Erkenntnis II*, pages 313–33, 1932. Reprinted in A. Danto and S. Morgenbesser, editors, *Philosophy of Science*, pages 121–140. New York: New American Library.

G. Nerlich. How Euclidean geometry has misled metaphysics. *Journal of Philosophy*, 88(4):169–89, 1991. ISSN 0022362X. www.jstor.org/stable/2026946.

C. Peacocke. *The Primacy of Metaphysics*. Oxford: Oxford University Press, 2019.

Z. R. Perry. Properly extensive quantities. *Philosophy of Science*, 82(5):833–44, 2015. https://doi.org/10.1086/683323.

Z. R. Perry. *Physical Quantities: Mereology and Dynamics*. PhD thesis, 2016.

Z. R. Perry. On mereology and metricality. *Philosophers' Imprint*, forthcoming.

W. Quine. *Word and Object*. Cambridge, MA: MIT Press, 1960.

J. Read and T. Møller-Nielsen. Redundant epistemic symmetries. *Studies in History and Philosophy of Science Part B: Studies in History and Philosophy of Modern Physics*, 70:88–97, 2020. ISSN 1355-2198. https://doi.org/10.1016/j.shpsb.2020.03.002. www.sciencedirect.com/science/article/pii/S1355219819301649.

F. Roberts. *Measurement Theory*, volume 7 of *Encyclopedia of Mathematics and Its Applications*. Reading, MA: Addison Wesley, 1979.

J. T. Roberts. A case for comparativism about physical quantities. Paper presented to the second annual conference of the Society for the Metaphysics of Science, Geneva, 2016. www.academia.edu/28548115/A_Case_for_Comparativism_about_Physical_Quantities_–_SMS_2016_Geneva.

B. Russell. *The Principles of Mathematics*. New York: W. W. Norton & Company, 1903.

S. Saunders. Mirroring as an a priori symmetry. *Philosophy of Science*, 74:452–80, 2007.

J. Schaffer. What not to multiply without necessity. *Australasian Journal of Philosophy*, 93(4): 644–64, 2015. https://doi.org/10.1080/00048402.2014.992447.

C. T. Sebens. Electron charge density: A clue from quantum chemistry for quantum foundations. *Foundations of Physics*, 51(75), 2021. doi: https://doi.org/10.1007/s10701-021-00480-7.

D. Wallace. Observability, Redundancy, and Modality for Dynamical Symmetry Transformations. In J. Read and N. J. Teh, editors, *The Philosophy and Physics of Noether's Theorems: A Centenary Volume*, 322–53. Cambridge: Cambridge University Press, 2022. https://doi.org/10.1017/9781108665445.014.

H. Weyl. *Philosophy of Mathematics and Natural Science*. Princeton, NJ: Princeton University Press, 1949.

J. Wilson. Fundamental determinables. *Philosopher's Imprint*, 12(4):1–17, 2012.

J. Wolff. *The Metaphysics of Quantities*. Oxford: Oxford University Press, 2020.

Acknowledgements

I would like to thank Guido Bacciagaluppi, David Baker, Harvey Brown, Adam Caulton, Eddy Keming Chen, Erik Curiel, Shamik Dasgupta, Marco Dees, Neil Dewar, Jeroen van Dongen, Patrick Dürr, Jamee Elder, Sam Fletcher, Caspar Jacobs, Mahmoud Jalloh, Dennis Lehmkuhl, Niels Linnemann, Céline Martens, Edith Martens, Lars Martens, Toine Martens, Casey McCoy, Tushar Menon, Thomas Møller-Nielsen, Zee Perry, Oliver Pooley, Carina Prunkl, James Read, John Roberts, Simon Saunders, Ted Sider, Syman Stevens, Reinier van Straten, Chris Timpson, Teru Tomas, Milica Veličković, David Wallace, Alastair Wilson, Jo Wolff and Chris Wüthrich for invaluable support, discussions and/or feedback, and James Owen Weatherall for his support as the editor of this Element.

Cambridge Elements

The Philosophy of Physics

James Owen Weatherall
University of California, Irvine

James Owen Weatherall is Professor of Logic and Philosophy of Science at the University of California, Irvine. He is the author, with Cailin O'Connor, of *The Misinformation Age: How False Beliefs Spread* (Yale, 2019), which was selected as a *New York Times* Editors' Choice and Recommended Reading by *Scientific American*. His previous books were *Void: The Strange Physics of Nothing* (Yale, 2016) and the *New York Times* bestseller *The Physics of Wall Street: A Brief History of Predicting the Unpredictable* (Houghton Mifflin Harcourt, 2013). He has published approximately fifty peer-reviewed research articles in journals in leading physics and philosophy of science journals and has delivered over 100 invited academic talks and public lectures.

About the Series

This Cambridge Elements series provides concise and structured introductions to all the central topics in the philosophy of physics. The Elements in the series are written by distinguished senior scholars and bright junior scholars with relevant expertise, producing balanced, comprehensive coverage of multiple perspectives in the philosophy of physics.

Cambridge Elements

The Philosophy of Physics

Elements in the Series

Foundations of Quantum Mechanics
Emily Adlam

Physics and Computation
Armond Duwell

Epistemology of Experimental Physics
Nora Mills Boyd

Structure and Equivalence
Neil Dewar

Emergence and Reduction in Physics
Patricia Palacios

The Classical–Quantum Correspondence
Benjamin H. Feintzeig

Idealizations in Physics
Elay Shech

The Temporal Asymmetry of Causation
Alison Fernandes

Special Relativity
James Read

Philosophy of Particle Physics
Porter Williams

Foundations of Statistical Mechanics
Roman Frigg and Charlotte Werndl

Philosophy of Physical Magnitudes
Niels C. M. Martens

A full series listing is available at: www.cambridge.org/EPPH

www.ingramcontent.com/pod-product-compliance
Ingram Content Group UK Ltd.
Pitfield, Milton Keynes, MK11 3LW, UK
UKHW020405180125
453697UK00007B/146